TECHNICAL
REPORT

A Review of the Army's Modular Force Structure

Stuart E. Johnson, John E. Peters,
Karin E. Kitchens, Aaron Martin

Prepared for the Office of the Secretary of Defense

RAND NATIONAL DEFENSE RESEARCH INSTITUTE

The research described in this report was prepared for the Office of the Secretary of Defense (OSD). The research was conducted within the RAND National Defense Research Institute, a federally funded research and development center sponsored by OSD, the Joint Staff, the Unified Combatant Commands, the Navy, the Marine Corps, the defense agencies, and the defense Intelligence Community under Contract W74V8H-06-C-0002.

Library of Congress Control Number: 2011930376

ISBN: 978-0-8330-5130-1

The RAND Corporation is a nonprofit institution that helps improve policy and decisionmaking through research and analysis. RAND's publications do not necessarily reflect the opinions of its research clients and sponsors.

RAND® is a registered trademark.

Published 2011 by the RAND Corporation
1776 Main Street, P.O. Box 2138, Santa Monica, CA 90407-2138
1200 South Hayes Street, Arlington, VA 22202-5050
4570 Fifth Avenue, Suite 600, Pittsburgh, PA 15213-2665
RAND URL: http://www.rand.org/
To order RAND documents or to obtain additional information, contact
Distribution Services: Telephone: (310) 451-7002;
Fax: (310) 451-6915; Email: order@rand.org

Preface

In 2003, the U.S. Army began implementing a set of ambitious changes to its force structure to address the challenges of waging war and conducting extended stabilization operations. It has done this while engaged in a multiple-theater war. One of the changes involved transforming the Army from its traditional, division-based force into a brigade-based force, a concept that has come to be known as "modularity." Although it was the proximate focus of this study, this move was not made in isolation. It was accompanied by two other force structure change initiatives and a major force management change initiative, all of which were roughly concurrent. The first force structure change was to "grow the Army" by raising end strength, thereby allowing the Army to add units. The second was to rebalance the force, moving some supporting capabilities from the reserve component to the active component. The aim was to bring the reserve component force structure into closer alignment with that of the active component. This rebalancing also moved more manpower into the tactical part of the Army and reduced the size of the institutional Army—the sustaining base—to an unprecedented low percentage of the total Army. The major force change initiative occurred in 2006, when the Army Force Generation process moved the Army from a tiered readiness to a cyclical readiness model. Given the near-simultaneity of these events, it is not surprising that their effects have become inextricably entangled, limiting the ability to isolate cause from effect. Congress directed this study to determine whether, by converting to a modular force, the Army has improved its capabilities.

In the move to the brigade-centric force structure, or modularity, the Army replaced its division-centric force structure with a force whose constituent building blocks are brigades and brigade combat teams (BCTs). BCTs were rebuilt by making proportionate combat, combat support, and combat service support, formerly provided by the host division, organic to the BCTs' organization. In the process, the Army reduced the number of combat brigade types in its force structure, from some 17 individual types to three: infantry BCTs, heavy BCTs, and Stryker BCTs. The move to modularity provided the Army with a greater number of smaller, very capable force packages, making it easier to sustain the protracted operations in Iraq and Afghanistan. Combat support and combat service support units and force structure were also redesigned to make the entire force more modular. However, the focus of this report is on the combat arms portion of the force structure and its operational command-and-control capabilities (or operational headquarters capabilities).

The research described in this report was sponsored by the Office of the Secretary of Defense, Capability Assessment and Program Evaluation, and conducted within the International Security and Defense Policy Center of the RAND National Defense Research Institute, a federally funded research and development center sponsored by the Office of the Secretary of

Defense, the Joint Staff, the Unified Combatant Commands, the Navy, the Marine Corps, the defense agencies, and the defense Intelligence Community.

For more information on the RAND International Security and Defense Policy Center, see http://www.rand.org/nsrd/about/isdp.html or contact the director (contact information is provided on the web page).

Contents

Figures

Tables

Summary

Congress has taken great interest in the capabilities of the U.S. Army as the service has transitioned to its new, modular force structure. Congress recently requested an independent study to determine

(A) The operational capability of the Army to execute the core mission of the Army to contribute land power to joint operations.

(B) The ability to manage the flexibility and versatility of Army forces across the range of military operations.

(C) The tactical, operational, and strategic risk associated with the heavy, medium, and light modular combat brigades and functional support and sustainment brigades.

(D) The required and planned end strength of the Army.[1]

This study addressed these questions through comparative analysis, considering how the earlier, division-centric force structure compared to the current force structure. Many of the data to support the analysis were provided by the Army. In particular, the U.S. Army Training and Doctrine Command's Army Capabilities Integration Center provided Army Structure Messages that allowed us to track changes in the Army's force structure between 2003 and 2008, the period when much of the transition to the current modular force structure was taking place. The research team also made use of the Army's Structure and Manpower Allocation System (SAMAS) database and FMSWeb for data on manpower and organizational characteristics of the force.

We concluded that the present force structure is superior to the earlier force structure in terms of its ability to contribute land power to current and reasonably foreseeable joint operations, its flexibility and versatility across the range of military operations, and its associated risks. The analysis also demonstrated that the modular force structure produced a larger tactical force with a larger number of more aggregated capabilities than its predecessor force structure.

Several insights from the analysis are worth special mention here. First, the Army's long-standing practice of task organization (the process whereby senior commanders temporarily attach some of their units to subordinate commanders to execute a given mission) is not a symptom of flaws in the Army's organizational design. Rather, it is a practice that allows a commander to tailor the force to the specific requirements of a given operation. For example, a commander might strip units from a subordinate commander who has a relatively easier

---

[1] Public Law 111-84, National Defense Authorization Act for Fiscal Year 2010, October 28, 2009, Sec. 344.

task in the operation and assign them to another subordinate who faces more demanding tasks. Task organization is a practice that contributes to the Army's flexibility and versatility, allowing it to adapt to changing circumstances.

Second, larger headquarters at the BCT, division, and corps levels are not in themselves signs of bloat. They too contribute to the force's flexibility and versatility because their size allows them to operate around the clock. They have more senior noncommissioned officers (NCOs) and field-grade officers than was the case in earlier headquarters designs, which enables better planning and wider spans of control (that is, the headquarters can effectively employ and control a larger number of subordinate maneuver units and integrate enabling organizations). Furthermore, senior NCOs and officers can be deployed to command company teams, battalion task forces, and other ad hoc organizations that allow the BCT, division, or corps to respond to unexpected challenges (or opportunities).

Third, the current force structure has been developed for current operations. It is very different in design from the forces that the Army organized to oppose the Soviet Western Group of Forces, but that adversary no longer exists. The current force structure provides "overmatch" (overwhelming advantages) against today's foes. Moreover, the process of adapting and adjusting the Army's fielded forces continues. In this regard, the move to the current force structure should not be viewed as revolutionary; it really reflects long-standing Army practices of reacting to changed circumstances by updating its forces and their organization, doctrine, and equipment.

Finally, concerns have been expressed regarding the absence of a third maneuver battalion in the heavy and infantry BCTs. The Army has been aware, almost from the beginning of the move to the brigade-centric force, of the limitations that the missing maneuver battalion imposes on the force, and it has crafted compensating tactics, techniques, and procedures. Former BCT commanders with whom we spoke would surely prefer a third maneuver battalion, but none believed that the two-battalion organization has led to greater risk in current operations. If circumstances changed and more capable adversaries appeared, the Army could reorganize to provide a smaller number of larger, three-battalion BCTs with the same end strength if it concluded that doing so would provide a clear advantage.

Acknowledgments

We are indebted to many people for their help over the course of this study. In the Office of the Secretary of Defense, Capability Assessment and Program Evaluation, we are grateful to our sponsor, Kathleen Conley, director, land forces, for her guidance and encouragement. We also thank John R. Duke, who served as the day-to-day facilitator for the research, organizing data requests from the Army, coordinating briefings, and capably handling the administrative chores associated with a project of this scope. In the Army, we are indebted to Dean Pfoltzer for his efforts in orchestrating Army participation in the research effort. Johnny Thomas, Headquarters, U.S. Department of the Army, Office of the Deputy Chief of Staff for Force Development (G-8), was also instrumental in coordinating Army-side logistics for the study, securing briefing rooms, and coordinating the schedules of busy Army stakeholders. Rickey Smith, U.S. Army Training and Doctrine Command, Army Capabilities Integration Center (Forward), provided valuable insight into the series of decisions that led to the present modular force program. We thank Matthew Markel at RAND and Terry Pudas at the National Defense University for their thorough reviews of this report and for their thoughtful suggestions, which were helpful in ensuring the rigor and relevance of this research effort. At RAND, we thank our colleagues in the Army research division, RAND Arroyo Center, Jeffrey Isaacson, Timothy Bonds, Bruce Held, Laurinda Rohn, and Thomas Szayna for their counsel and insights. We also thank our program director, James Dobbins, for his oversight and helpful engagement throughout this research effort.

Abbreviations

AC	active component
ARCIC	Army Capabilities Integration Center
ARFORGEN	Army Force Generation
ARSTRUC	Army structure
BCT	brigade combat team
BOG	boots on the ground
COCOM	combatant command
CS	combat support
CSA	U.S. Army Chief of Staff
CSS	combat service support
DISCOM	division support command
DoD	U.S. Department of Defense
FY	fiscal year
GF	generating force
HQDA	Headquarters, U.S. Department of the Army
JTF	joint task force
METT-TC	mission, enemy, terrain and weather, troops and support available–time available, and civil considerations
MOS	military occupational specialty
MP	military police
MTO&E	Modified Table of Organization and Equipment
NCO	noncommissioned officer
OEF	Operation Enduring Freedom
OF	operating force

OIF	Operation Iraqi Freedom
PSYOPS	psychological operations
QDR	Quadrennial Defense Review
RC	reserve component
RSTA	reconnaissance, surveillance, and target acquisition
TDA	Table of Distribution and Allowances
TOW	tube-launched, optically tracked, wire-guided
TRADOC	U.S Army Training and Doctrine Command
UA	unit of action

CHAPTER ONE

Introduction

Army modularity—restructuring the force to produce a supply of directly interchangeable units—is the product of a number of experiences and concerns.[1] Modularity may be best characterized as the Army's institutional response to a host of factors, some stretching back to the early 1990s, that caused the Army to move away from its traditional division-based force to a brigade-centric design. The most salient considerations included the following:

- There was a realization (circa 1990) that the Cold War was over and that, in its absence, the United States' standing was no longer endangered by a revisionist power with direct military force. In brief, there was no "peer competitor."
- There were concerns about the Army's enduring national security relevance after 1999's Operation Allied Force succeeded against Milosevic's Serbia without recourse to ground forces, as well as the Army's perceived difficulty in deploying its Task Force Hawk in support of the allied campaign there.[2]
- There was a subsequent desire by then–U.S. Army Chief of Staff (CSA, 1999–2003) General Eric K. Shinseki for a responsive, mobile, midweight (that is, mobile with its own vehicles) force that would be deployable by air to any crisis in the world within 96 hours.
- The terrorist attacks of September 11, 2001, increased demand for Army forces.
- The Army became engaged in extended campaigns in Afghanistan and Iraq.
- The Army was required to support enduring global counterterrorism efforts.
- The Quadrennial Defense Review (QDR) and other defense guidance emphasized building partnerships and partner capacity and the provision of security force assistance to beleaguered friendly states around the world.
- There was a growing requirement to support civil authorities in homeland defense and domestic disaster mitigation and response.
- The Army faced the imperative to be flexible and adaptive in providing appropriate military capabilities to combatant commanders as the United States prosecuted its campaigns against a variety of adversaries and antagonists.

---

[1] The process also involved the creation of "functional support" and "multifunctional" brigades to provide key types of support.

[2] In fact, Task Force Hawk met all its timelines for deployment. See Bruce Nardulli, Walter L. Perry, Bruce R. Pirnie, John Gordon IV, and John G. McGinn, *Disjointed War: Military Operations in Kosovo, 1999*, Santa Monica, Calif.: RAND Corporation, MR-1406-A, 2002, Chapter Four.

Perhaps most significantly, according to Army officials consulted in the course of this research,[3] the modular force structure and the Army Force Generation (ARFORGEN) process have together been central to the Army's ability to provide a steady stream of forces to satisfy the demands imposed by operations in Afghanistan and Iraq.

Congress has raised four specific questions about the modular force and placed language in Section 344 of the National Defense Authorization Act for Fiscal Year 2010, Public Law 111-84, that prompted this study. Specifically, Congress directed a study on Army modularity to determine

(A) The operational capability of the Army to execute the core mission of the Army to contribute land power to joint operations.

(B) The ability to manage the flexibility and versatility of Army forces across the range of military operations.

(C) The tactical, operational, and strategic risk associated with the heavy, medium, and light modular combat brigades and functional support and sustainment brigades.

(D) The required and planned end strength of the Army.

This report describes the methods and results of the congressionally mandated study. The remainder of this chapter outlines the study's methods, data sources, and analysis. Chapter Two discusses the Army's journey from a division-based force to the current modular, brigade-centric force. In doing so, it summarizes the considerations motivating the Army's senior leaders and those in the Office of the Secretary of Defense to take the Army through this transformative process. Chapter Two also describes the structural effects of modularity on the Army, highlighting the organizational changes that resulted. Readers should leave Chapter Two with a clear appreciation of the structural differences between the premodular, division-based force and its modular successor and the implications for the Army's ability to satisfy demand for its forces.

Chapters Three, Four, Five, and Six, respectively, address the four issues posed by the Congress. The chapters do not closely follow a single template, though they all share a comparative approach: considering the modular force's ability relative to that of the previous, division-based structure. An appendix addresses a series of additional considerations outlined by Congress.

Study Approach

The research team began by analyzing each of the questions posed by Congress. This process included interviews with the congressional staffers responsible for the original language in the National Defense Authorization Act to ensure that the team understood the concerns. The research team also prepared a preliminary research outline based on the congressional language and deduced from it the types of data and information required to address each issue. These data and information requirements became a "data call," or request for information,

---

[3]   This study benefited from a study advisory group chaired by officials from the Office of the Secretary of Defense, Capability Assessment and Program Evaluation, and populated by representatives from the Army Staff, including the G-3/5/7 and G-8.

which the study advisory group forwarded to the appropriate parts of the Army. The study team also delivered a set of questions to the Army through the study advisory group. When the questions were answered, the research team had sufficient information to come to a judgment on each issue in the congressional mandate. At that point, the team reviewed the answers to the individual questions and developed a conclusion on each overarching issue.

The next section describes the research approach used to address each issue and how the research team reached the specific determinations that Congress sought.

The Army's Capability to Contribute Land Power in Joint Operations

This issue was examined comparatively. Modified Tables of Organization and Equipment (MTO&Es) represent a summary of unit capabilities: the number of troops assigned, their respective skills in combat, and combat support (CS) and combat service support (CSS) military occupational specialties (MOSs). MTO&Es also reflect the unit's equipment: vehicles, weapons, command-and-control and communication capabilities, intelligence and reconnaissance systems, and so on. Thus, they provide a reasonable basis for comparing modular units' capabilities with those of their antecedents in the division-based Army. We expanded on this MTO&E comparison to calculate the units' relative ability to control territory and to protect indigenous populations by calculating their force-to-space and force-to-population ratios. We made these comparisons for light, airborne, and heavy formations. In the case of the premodular force, we aligned a combat brigade's nonorganic but habitually associated supporting forces from the division base so that comparisons with today's brigade combat teams (BCTs) would reflect the full suite of combat, CS, and CSS available to both the divisional brigade and the modular force BCT.

We also analyzed the relative density of combat, CS, and CSS resources in the units. We calculated "tooth-to-tail" ratios and similar measures of merit to identify differences in the organization of these units. We extended these comparisons to a macro level, analyzing the ability of each force structure to produce balanced force packages through the ARFORGEN process; this helped us determine where the current force structure might run out of enabler units (e.g., sustainment brigades, mobility enhancement brigades). Finally, we used Army Structure (ARSTRUC) Messages to track changes in the overall force structure, allowing us to identify the units that the Army had disbanded to achieve the current force structure.

The net result of these comparisons was an ability to see how today's Army differs structurally from its predecessor, what those structural differences indicate about the relative capabilities of the earlier and current force structure, and, thus, each structure's ability to contribute land power to joint operations.

Options for Managing the Flexibility and Versatility of Army Forces Across the Range of Military Operations

Our hypothesis was that the capacity to manage flexibility and versatility resides within Army unit headquarters, specifically those of the brigade or BCT, the division, and the corps: The more capable the headquarters, the more force flexibility and versatility it commands. Flexibility and versatility are force structure attributes that reflect (1) the relative abundance of combat, CS, and CSS in the force structure; (2) the ease with which these resources can be packaged together to support a commander or an operation; and (3) the amount of turbulence caused by task organization to provide the necessary capabilities.

To measure the relative capabilities of a headquarters, we conducted direct MTO&E comparisons of brigade and BCT headquarters and of pre-modular and modular division and corps headquarters. To measure the relative abundance of forces, we used the MTO&E analysis undertaken for the first congressional question to determine the shift in forces from the reserve component (RC) to the active component (AC) and the resulting abundance of combat, CS, and CSS capabilities in each component. To measure the ease of packaging and the amount of turbulence caused by task organization, we counted units rendered less effective by the removal of key subelements to task-organize necessary force packages. Based on these comparisons, we arrived at conclusions about the relative ability to manage flexibility and versatility of Army forces and the relative amounts of flexibility and versatility within those forces.

Tactical, Operational, and Strategic Risks Associated with Brigade Combat Teams, Multifunctional Support Brigades, and Functional Brigades

Our approach to risk was also comparative: risk encountered by the premodular force relative to risk confronted by today's force. Our conception of risk was classical, the expected value, where risk equals the probability of events occurring multiplied by their consequences. We conceived of risk in these terms at the strategic, operational, and tactical levels of warfare to investigate two circumstances: the appearance of a very capable adversary prepared to confront Army forces in a classical, Clausewitzian force-on-force contest and irregular warfare or insurgency that evolves to such a scale as to be beyond the capacity of the Army to manage it. Understood in these terms, risk depends on the Army's ability to satisfy demand for its forces, either to defeat the newly emerged adversary or to defeat the insurgency.

We also conducted interviews with former BCT commanders to capture their views on risk in current operations. Specifically, we asked them whether they believed that the current force structure—compared to its predecessor—reduced the risk to Army units, or whether risk remained about the same or even increased. By combining the answers from these interviews with our analysis of risk emanating from capable adversaries or large insurgencies, we reached judgments about the strategic, operational, and tactical risk confronting today's force structure.

Required and Planned End Strength of the Modular Force

We sought to determine whether the move to modularity resulted in additional demands on end strength beyond those arising from ongoing military operations. To this end, we analyzed the timing and size of Army end-strength increases, the timing of the transition to the modular force structure, and the fungibility of Army manpower to fill MOS-particular duty assignments in the modular MTO&E force structure. Our analysis drew significantly on the data we had gathered and processed for the earlier tasks in the study.

Additional Considerations

Congress posed some *additional considerations* to be addressed by the study. It asked that the study examine the Army's historical experience with separate brigade structures, the original Army analysis upon which the modular force designs were based, and subsequent analyses that confirmed or modified the original designs. Congress also asked that the study consider lessons learned from Operation Enduring Freedom (OEF) and Operation Iraqi Freedom (OIF) and, in particular, how modular formations were task-organized and employed in ways that may have differed from the original modular concept or that may have confirmed or modified the original designs. Congress also expressed an interest in understanding improvements made in implementing brigade and headquarters designs and in the deployability, employability, and sustainability of modular formations compared with the earlier, premodular designs.

This report addresses these considerations in the course of Chapters Two and Three, with greater detail provided in an appendix. Chapter Two describes the Army's analysis leading to modular brigades and the Army's earlier experience with separate brigades and lessons from OIF and OEF. Chapter Three analyzes changes in brigade and headquarters designs and the relative deployability, employability, and sustainability of modular forces compared with predecessor formations.

Data Sources

The principal types of data sources used and the utility of these sources fall into the following six broad categories:

- congressional testimony by incumbent Army Chiefs of Staff revealing the concerns, preferences, and motives leading them to favor actions that ultimately produced the modular Army
- non–U.S. Department of Defense (DoD) reports, typically from the Congressional Budget Office, Congressional Research Service, and Government Accountability Office, that analyze and report on aspects of or issues connected with the transition to the modular force
- statements by officials in the Office of the Secretary of Defense and the Army identifying positive and negative attributes or the performance of Army forces, both modular and premodular
- official Army documents (e.g., TO&E) documenting specific characteristics of Army units, including staffing levels, equipment, and capabilities.
- internal Army documents, such as briefings, information papers, internal histories, after-action reports, and surveys of Army tactical leaders (e.g., BCT commanders) characterizing the performance of units, including their positive and negative attributes and constraints on their effectiveness
- Army proprietary websites, particularly the U.S. Army Force Management Support Agency's FMSWeb.

Methods of Inquiry

We addressed the questions from Congress by comparing the modular Army's capabilities and attributes with those of the premodular Army. To the maximum extent possible, we relied on quantitative analysis. Where quantitative information was not available, we used qualitative information derived from the data sources. In this regard, we sought expert opinion and official judgments about the performance of Army formations found in the pages of the documents cited in the bibliography.

The project also hosted interim reviews with stakeholders from the Office of the Secretary of Defense and the Army. At these reviews, we presented our preliminary findings and invited the stakeholders to critique the research, suggest additional analyses that might add clarity or correct a misapprehension, and offer their own views on what the conclusions mean in terms of the congressional questions. We subsequently incorporated these views into the study where it was appropriate to do so.

We also sought the views of subject-matter experts, especially former brigade commanders who commanded in combat. Our goal in adopting this approach was to produce robust, transparent, and replicable answers to the congressional questions about the modular force.

The following chapter begins by reviewing the considerations that caused the Army's leaders to move away from the previous force structure design and to embrace modularity.

CHAPTER TWO

The Impetus for Modularity

The history of "modularity" began in earnest in 2003 with CSA General Peter J. Schoomaker's idea to create a more effective fighting force by moving the Army from a division-based to a brigade-based structure. This concept built on reflections on the Army's emerging role with the close of the Cold War. As that period came to an end, the Army recognized that change was needed to face the realities of the new security environment. This recognition began the transformation process. The Army Chiefs of Staff, from General Gordon R. Sullivan to General Schoomaker, all recognized the need to adapt. Modularity did not come about in isolation. It was part of a process that began in the early 1990s. Modularity constituted an increasingly important component of efforts to transform the Army, and it became the principal component of Army transformation under General Schoomaker. Each CSA since then has added his own vision to the transformation process.

General Sullivan took many of the first steps toward transforming the Army in the early 1990s. He realized that the Army needed to adapt even as political pressures for a "peace dividend" drove the service's end strength downward. The U.S. Army Training and Doctrine Command (TRADOC) echoed this need and published Pamphlet 525-5, *Force XXI Operations*, in August 1994. It contained one of the first references to modularity. The theme of the pamphlet was change: Change would come in the way the Army organizes, trains, mobilizes, projects, and sustains the force. The scale and pace of shifts in the global security environment required an Army that could adjust quickly: "Change will continue, requiring our Army to recognize it as the only real constant."[1] This ever-altering environment would require adaptive and flexible forces. Modularity was identified as one of the means to achieve these goals. TRADOC Pamphlet 525-5 defined modularity as "a force design methodology that establishes a means to provide interchangeable, expandable, and tailorable force elements."[2]

In 1993, the Army had a change in doctrine, outlined in Field Manual 100-5, which stated that the Army must be prepared for a range of full-dimensional operations in war to operations other than war.[3] TRADOC Pamphlet 525-5 built on this doctrinal shift. The Army must shift from the "deterministic and very appropriate scientific approach of the Cold War" to "emphasizing a concept built on principles that must be translated to action in specific scenarios that cannot now be predicted with enough certainty to warrant a return to pre-

---

[1] U.S. Army Training and Doctrine Command, *Force XXI Operations*, Pamphlet 525-5, Fort Monroe, Va., August 1, 1994, para. 1-1.

[2] TRADOC, 1994, Glossary, p. 5.

[3] Headquarters, U.S. Department of the Army, *Operations*, Field Manual 100-5, Washington, D.C., June 14, 1993, p. vi.

scriptive doctrine."[4] Other factors were driving forces as well, including shifting and unstable balances of power in the Balkans, the Middle East, and throughout large parts of Africa and Asia. Information technology also drove the transformation. Technology greatly increased the speed, volume, and accuracy of information being received. Modularity challenged the Army's traditional hierarchical structure of battle command, however. Although the implications of nonhierarchical models were not fully understood, it was believed that the coexistence of both hierarchical and nonhierarchical processes "will result in military units being able to decide and act at a tempo enemies simply cannot equal."[5]

TRADOC Pamphlet 525-5 set the stage for Force XXI, which was defined by five characteristics:

- doctrinal flexibility
- strategic mobility
- tailorability and modularity
- joint and multinational connectivity
- versatility to function in war and operations other than war.[6]

"The Nation cannot afford to maintain an *army of armies* in the early twenty-first century."[7] The Army must therefore be prepared for a diverse set of circumstances.

Modularity and tailorability went hand in hand. The idea was that organizations tend to "grow flatter and less rigidly hierarchical" in response to increases in technology. Versatility would improve if the Army could create forces that were "as modular as logic allows," so they could be tailored to meet each contingency.[8] Force XXI would be "smaller, yet have new, expanded, and diverse missions in an unpredictable, rapidly changing world environment."[9] Although the division remained the primary tactical formation, the roots of modularity are evident in Force XXI.

TRADOC 525-68 was released in 1995 and delved more deeply into the concept of modularity. The end of the Cold War reduced the imperative of forward presence and rapid reinforcement for the Army and shifted the emphasis to force projection. The range of missions in which the Army was engaging expanded and diversified. TRADOC Pamphlet 525-68 recognized the difficulty of sending the appropriate-sized units to fit mission needs:

> Often times, commanders require a function to be performed which does not warrant the deployment of an entire unit. However, deploying portions of units can render the parent organization incapable of performing its mission.[10]

---

[4]   TRADOC, 1994, para. 1-3.

[5]   TRADOC, 1994, para. 1-2.

[6]   TRADOC, 1994, para. 3-1.

[7]   TRADOC, 1994, para. 3-2; emphasis added.

[8]   TRADOC, 1994, para. 3-2.

[9]   TRADOC, 1994, para. 4-1.

[10]   U.S. Army Training and Doctrine Command, *Concepts of Modularity*, Pamphlet 525-68, Fort Monroe, Va., January 10, 1995, Forward.

One way to address this problem was modularity, which would improve the Army's ability to respond to a diverse spectrum of operations while deploying the minimum number of troops and equipment adequate to do the job. Modularity, at the very beginning, was about deploying "the right amount of the right functions and capabilities in the right place at the right time."[11]

The motivations for this new concept centered on the shift to prepare for more diverse, less established theaters that required different types of Army forces. Task organization is "a temporary grouping of forces designed to accomplish a particular mission."[12] In these scenarios, old methods of task organizing did not optimize capabilities; they often involved deploying only pieces of an organization (typically a division), rendering the remaining portion unbalanced and incapable of performing its missions. The new threats, the Army perceived, would require "more efficient packaging of force capability which can be provided through modularity."[13]

General Dennis J. Reimer continued the path forged by General Sullivan but expanded the Force XXI concept by creating the Army After Next, which targeted a new generation of weapon systems to be fielded by 2025. The goal of the Army After Next was to link Force XXI with the Army's long-term vision through a series of high-level war games.[14] The concept of the mobile strike force emerged from these games. This concept was to create a force that was more rapidly deployable than its heavy counterparts but also more survivable and lethal than the current light forces. Lessons learned in past operations added to the need for the "medium-weight" force. Operations Desert Storm and Joint Endeavor, in the view of the Army's senior leadership, both showed a lack of speed and mobility in deploying heavy forces. Simulations suggested that the basic combat vehicle for the strike force needed to be wheeled for the future battlefield—something that the Army did not currently have. There would be three stages to creating the strike force: an entry stage, and interim stage, and an objective stage. The entry and interim stages consisted of simulations, testing, and validation. The objective stage, during which implementation of the force would occur, was set to begin in 2005.[15]

When General Eric K. Shinseki became Chief of Staff in 1999, he continued the initiatives begun by his two predecessors but believed change should occur more rapidly. He developed the new Army vision: "Soldiers on point for the Nation transforming the most respected Army into a strategically responsive force that is dominate across the full spectrum of operations."[16] The intent was to create an Army that was "more deployable, more agile, more versatile, more lethal, more survivable, and more sustainable."[17] A key goal was for the Army to be able to field a combat-ready brigade anywhere in the world within 96 hours, a division

---

[11] TRADOC, 1995, Forward.

[12] Headquarters, U.S. Department of the Army, and U.S. Marine Corps, *Operations Terms and Symbols*, Field Manual 101-5-1 and Marine Corps Reference Publication 5-2A, Washington D.C., September 30, 1997.

[13] TRADOC, 1995, p. 2.

[14] Walter L. Perry and Marc Dean Millot, *Issues from the 1997 Army After Next Winter Wargame*, Santa Monica, Calif.: RAND Corporation, MR-988-A, 1998.

[15] Mark G. Cianciolo, *U.S. Army Strike Force—A Relevant Concept?* Ft. Leavenworth, Kan.: School of Advanced Military Studies, April 1999, pp. 12–13.

[16] General Eric K. Shinseki, U.S. Army Chief of Staff, prepared statement before the Senate Armed Services Committee, October 26, 1999.

[17] Shinseki, 1999.

within 120 hours, and five divisions within 30 days.[18] The force structure of the day meant that the force was not very mobile or survivable.

The transformation would occur on three major paths: the legacy force, the interim force, and the objective force. One of the major components of the interim force would be the Stryker brigade that had been initiated by General Reimer.[19] General Shinseki established the 3rd Brigade of the 2nd Infantry Division to be the prototype of the new medium-weight force. The goal was to have the prototype ready by the end of fiscal year (FY) 2000. Lessons learned from the Army After Next and strike force were key to shaping the medium-weight force. The Stryker vehicle was chosen as the medium-weight armored vehicle and gave the name to the newly formed brigade: the Stryker brigade.[20] The Stryker brigade was the first explicitly modular brigade formation in any Army transformation process. It would be a test for the future of the Army and the "organizational structure, leader development, and training and soldier support concepts that could enable the future force imperatives of deployability, lethality, modularity, mobility, sustainability, and survivability that are required to address the new environment."[21]

Because of the terrorist attacks in September 2001, attention to developing the Stryker brigade intensified. It was believed that the program would be a critical step in equipping the Army for the Global War on Terrorism.[22]

The first Stryker brigade entered combat on December 3, 2003, during Operation Iraqi Freedom. As this convoy crossed into Iraq, the magnitude of the change became clear: The heavy equipment transport systems needed to carry Abrams tanks or Bradley fighting vehicles were unnecessary for the Stryker formations. The brigade was able to travel without the assistance of these systems. The success of the movement was reinforced by BCT commander Colonel Michael Rounds: "The 540-mile move speaks a lot about the capability of the brigade—our operational agility. We were relatively self-sustaining and self-moving."[23] This experience suggested that the best component to send as the building block of the Army would be a brigade that is adaptable and easy to deploy. The initial Stryker experience in Afghanistan was consistent with the Stryker brigade's yearlong tour in Iraq. The deployment to theater from the United States required only commercially chartered aircraft and two specialized sealift vessels. This demonstrated the Stryker unit's relatively responsive deployability. The unit also showed its flexibility by changing missions without requiring logistical assistance from higher headquarters.[24]

When General Schoomaker became CSA on August 1, 2003, he had a wealth of knowledge at his disposal: Force XXI, a portfolio of modeling and simulations from the Army After

---

[18] Mark J. Reardon and Jeffery A. Charlston, *From Transformation to Combat: The First Stryker Brigade at War*, Washington D.C.: U.S. Army Center of Military History, 2007, pp. 1–14.

[19] General Eric K. Shinseki, U.S. Army Chief of Staff, statement before the Senate Armed Services Committee, March 1, 2000.

[20] Reardon and Charlston, 2007, pp. 1–14.

[21] Wayne A. Green, *Interim Strike Force Headquarters Digital LNO Nodes: Force Tailoring Enablers*, Ft. Leavenworth, Kan.: Army Command and General Staff College, May 1999, p. 2.

[22] Reardon and Charlston, 2007, p. 14.

[23] Reardon and Charlston, 2007, p. 20.

[24] Reardon and Charlston, 2007, p. 69.

Next program, the Stryker brigade, and three previous CSAs who had the same end goal in mind as his, a versatile, deployable force capable of sustained operations against a range of adversaries. As General Schoomaker presented his plan to Congress, his testimony explained the imperatives driving the Army toward modularity. The Secretary of Defense's support aided General Schoomaker in prosecuting the rapid changes he sought to make.[25]

In a prepared statement before the House Armed Services Committee on July 21, 2004, General Schoomaker outlined the new realities that the Army and the United States faced:

> The single most significant component of our new strategic reality is that because of the centrality of the ideas in conflict, this war will be a protracted one. Whereas for most of our lives the default condition has been peace, now our default expectation must be conflict . . . a foreseeable future of extended conflict in which we can expect to fight every day, and in which real peace will be the anomaly.[26]

To sustain protracted operations in the ongoing wars and prevail over a variety of adaptive adversaries fighting on widely differing terrain under demanding circumstances, the Army would have to repackage its forces. Divisions were no longer the optimal unit of action. General Schoomaker believed that smaller units could better respond to challenges the Army was facing. "No single, large fixed formation can support the diverse requirements of full spectrum operations."[27] BCTs were determined to be the appropriate size. By shifting to BCTs, General Schoomaker believed that "we can significantly improve the tailorability, scalability, and 'fightablity' of the Army's contribution to the overall joint fight."[28] As for the current division-based task-organized force: "Tailoring and task-organizing our current force structure for such operations renders an ad hoc deployed force and a non-deployed residue of partially disassembled units, diminishing the effectiveness of both."[29] The brigades themselves needed improvement and were organized inefficiently; "right now, all these brigades are different—the number of helicopters in them, the number of units, sub-units within these brigades—and it's extraordinary inefficient."[30] Modularity would create a more efficient way of organizing a force with more standardized brigades, enabling direct interchangeability when it is necessary to replace a unit. In addition, modularity would increase the number of BCTs available through improved force management.[31]

General Schoomaker testified to Congress that modularity "will increase the active force by 30 percent with a minimal cost to our program. . . . That means that the dwell time issues, the kinds of pressures that are on the Guard, Reserve, and the active force are reduced." Gen-

---

[25] William M. Donnelly, *Transforming an Army at War: Designing the Modular Force, 1991–2005*, Washington, D.C.: U.S. Army Center of Military History, 2007, pp. 23–24.

[26] General Peter J. Schoomaker, U.S. Army Chief of Staff, prepared statement before the House Armed Service Committee, July 21, 2004b.

[27] Headquarters, U.S. Department of the Army, *Operations*, Field Manual 3-0, Washington, D.C., February 27, 2008, para. C-2.

[28] Schoomaker, 2004b.

[29] Schoomaker, 2004b.

[30] General Peter J. Schoomaker, U.S. Army Chief of Staff, testimony before the Senate Armed Services Committee, February 10, 2004a.

[31] Army employment data.

eral Schoomaker also believed that the Army was not properly utilizing the RC. A goal of modularity was to make National Guard and reserve units more readily able to integrate into the active force; "our initiatives of modularity that allow us to plug and play Reserve component units with active component units in a seamless fashion, of course, is our objective."[32]

The logistical structure also needed to be improved. "The Cold War Army designed its logistic structure for operations in developed theaters with access to an extensive host-nation infrastructure"; neither could be expected in future operations. To address this problem, General Schoomaker wanted to eliminate the layered support structure and bridge "the distance from theater or regional support commands to brigade combat teams with modular, distribution-based capabilities."[33]

Modularity would help maximize unit cohesion through habitual association among combat, CS, and CSS units, creating relationships of mutual confidence and loyalty within companies, battalions, and brigades, which would, in turn, make units more effective in combat.[34]

With conflict becoming the default condition, the individual staffing system was also proving to be inefficient. The level of readiness that the system could achieve was not sufficient for the repeated deployment and employment of major portions of the Army. Modularity would help by synchronizing soldiers' tours within their unit's rotation cycle and stabilizing the assignment of soldiers and their families at home stations and communities across recurring rotations.[35]

Another motivation of modularity was to improve joint operations by combining service capabilities. In the past, "the Army garrisoned the bulk of its tactical units to optimize economic efficiency and management convenience rather than combined-arms training." In other words, it "designed its capabilities to satisfy every tactical requirement autonomously, viewing sister service capabilities as supplementary." The Army needed to develop operational concepts, capabilities, and training programs "that are joint from the outset, not merely an afterthought." Modularity would aid in improving these joint capabilities. "The more modular the Army's capabilities, the better we will be able to support our sister services."[36]

The Modular Force and Today's Operational Concepts

In making the move to the modular force, the Army expected that its new force structure would have utility against a variety of adversaries in different types of terrain and that the new force structure would be suitable for both homeland defense and overseas contingency operations. Figure 2.1 contrasts expectations for the utility of the premodular force with expectations for the modular force. The top of the figure shows the premodular force, with its heavy brigades oriented toward open and mixed terrain and operations in or near urban terrain. The light force structure was likewise oriented on operations in or near urban terrain and on restric-

---

[32] Schoomaker, 2004a.

[33] Schoomaker, 2004b.

[34] Schoomaker, 2004b.

[35] Schoomaker, 2004b.

[36] Schoomaker, 2004b.

Figure 2.1
Comparative Expectations for the Utility of Army Forces

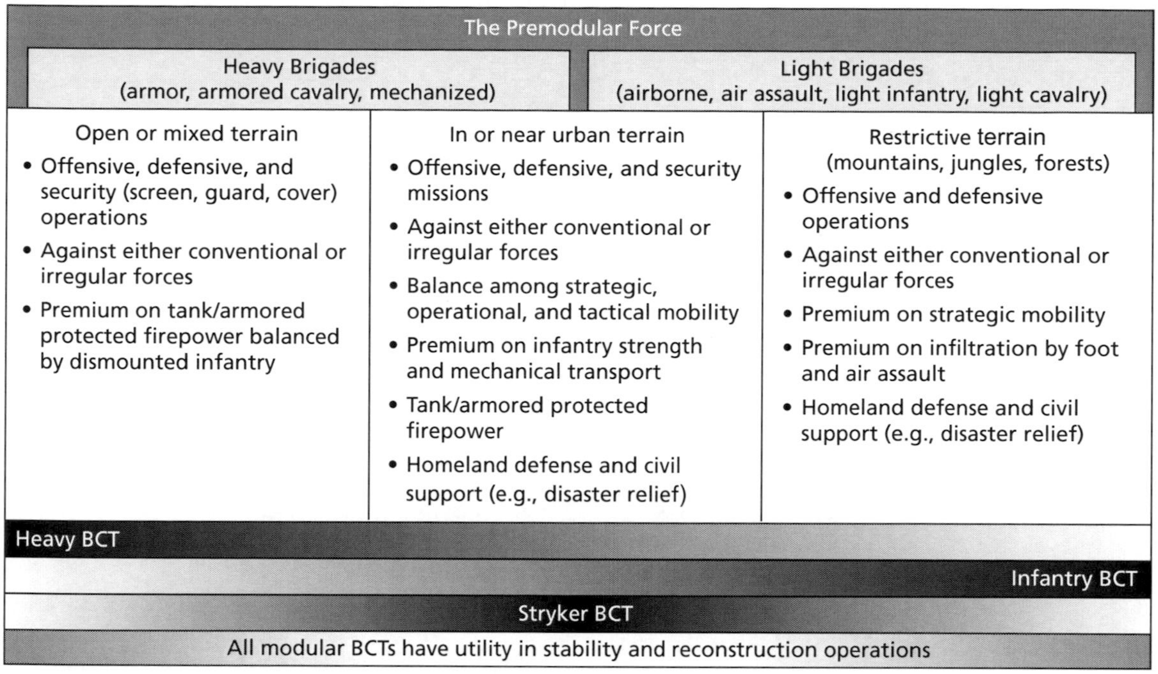

The Premodular Force		
Heavy Brigades (armor, armored cavalry, mechanized)		**Light Brigades** (airborne, air assault, light infantry, light cavalry)
Open or mixed terrain	In or near urban terrain	Restrictive terrain (mountains, jungles, forests)
• Offensive, defensive, and security (screen, guard, cover) operations • Against either conventional or irregular forces • Premium on tank/armored protected firepower balanced by dismounted infantry	• Offensive, defensive, and security missions • Against either conventional or irregular forces • Balance among strategic, operational, and tactical mobility • Premium on infantry strength and mechanical transport • Tank/armored protected firepower • Homeland defense and civil support (e.g., disaster relief)	• Offensive and defensive operations • Against either conventional or irregular forces • Premium on strategic mobility • Premium on infiltration by foot and air assault • Homeland defense and civil support (e.g., disaster relief)

Heavy BCT

Infantry BCT

Stryker BCT

All modular BCTs have utility in stability and reconstruction operations

SOURCE: Adapted from a briefing slide from HQDA, G-3/5/7, October 2004.
RAND TR927-2.1

tive terrain. The terrain panels in the figure reflect expectations about the types of operations and adversaries that Army units would face in each type of terrain, as well as the features of Army forces that would prove most useful there. The bands running across the bottom of the figure reflect the Army's expectations about the value of its current forces, given the enemy and terrain assumptions captured in the figure. The fact that the bands span the entire width of the figure suggests that the modular BCTs are designed to have utility across all the instances reflected in the figure, although the shading suggests that the greatest value for heavy BCTs lies in open and mixed terrain, while the greatest utility for infantry BCTs lies in restrictive terrain. Stryker BCTs present the greatest utility in or near urban terrain, according to the figure.

Other Key Assumptions

Figure 2.1 presents the Army's expectations for its new force structure at the tactical level. Others have captured the Army's strategic-operational expectations.[37] These expectations include

- a high degree of strategic mobility
- at least as much combat power as current heavy forces
- the likelihood that the "Army could probably never again expect to conduct major operations on its own"

[37] See Donnelly, 2007, pp. 7–9.

- that the force must be "versatile enough to deploy for almost any mission, from humanitarian assistance to major conventional war"
- that it "can work effectively with other American military services"
- that its "command elements might also have to serve as combined headquarters with the militaries of other nations or coordinate with nongovernmental agencies"
- that "information technologies . . . would allow fewer personnel to do as much or more than the larger staffs currently in place"
- that modularity would support "rapid, task-related configuration [of units] to do a specific job"
- that units would be "interchangeable and expandable [so as to be tailorable] to meet changing conditions"
- that such flexibility and versatility would be enabled by "effective information systems linked to reliable telecommunications [ensuring that] units involved in an operation were reliably connected"
- that "troops and their officers would sometimes assume greater responsibilities than those normally associated with their ranks or positions"
- that a modular force "might need more leaders of all ranks than a conventionally configured force" to "provide command and control to many independent elements that some missions would entail."[38]

Conclusions

The process of transforming the Army into the current, modular force has taken nearly two decades. The efforts of General Sullivan, General Reimer, and General Shinseki to define the future of the Army informed General Schoomaker's vision of the modular force. Each of these CSAs had the same goal: to move the Army from its Cold War posture toward a new posture that was better suited to the modern operations.

---

[38] Donnelly, 2007, pp. 7–9.

CHAPTER THREE

Determining the Army's Operational Capability to Contribute Land Power to Joint Operations

This chapter addresses the first issue outlined in the congressional mandate for this study, the Army's capability to contribute land power to joint operations.

Framing the Issue

There are two critical aspects to this issue. The first is obvious. The Army's ability to provide land power to military operations has been its core mission since the founding of the republic. As the security environment has changed, the Army has adapted its organization and doctrine to cope with the new challenges, sometimes in anticipation of changes on the battlefield, sometimes as a result of lessons learned on the battlefield. As relevant new technologies have been developed, the Army has strived to assimilate them into its forces to maintain an advantage over the enemy.

The second element is still being refined throughout the U.S. military: how to contribute forces configured for *joint* operations. There has always been an element of jointness in how the U.S. military fights, but attention to jointness intensified during the latter part of the Cold War. U.S. and allied ground forces were greatly outnumbered by those of the Warsaw Pact. To mitigate this imbalance, the Army worked with the Air Force to develop the Air-Land Battle concept, which exploited the capabilities of the two services to bring a coordinated, composite force to bear that was more effective than the sum of its individual parts.

Since that time, decisions about how to organize, train, and equip Army forces have, as a matter of course, been made in the context of joint operations. This process is here to stay. The United States has found that the intensification of combat power when the individual services plan with a joint operation in mind is compelling. This practice puts a premium on the Army's ability to coordinate its efforts on the battlefield with those of the other services.

The Problem

The demand signal for ground force capabilities, as the previous chapter noted, has changed. The key design consideration is to have a force structure and unit organization that permits the Army to respond quickly to crises that are frequent, often in austere theaters, of long duration, and well short of large-scale war with a peer competitor. In response to such circumstances, desirable force attributes include the following:

- rapidly deployable
- as capable as predecessor units
- agile
- tailorable
- more sustainable
- more survivable.

A key Army insight was that a division-based organization, although well suited for large deployments to engage in combat across a broad front, was less well suited for responding to the types of crises confronting the nation. For these threats, the Army's leadership concluded that it needed forces organized around units smaller than a division. The metaphor the leadership was fond of using was "I need a five dollar bill, but all I have in my wallet are twenties."

So the Army's leadership set out to create a force that could deploy in smaller packets without disrupting (or rendering ineffective) other parts of the force structure. At the same time, the force had to preserve the Army's capability to execute larger campaigns.

A Change in the Demand Signal

Since the terrorist attacks of 2001, the demand signal for land power has changed even more. While there is still a requirement for the capability (and capacity) to deter and defeat a highly capable regional aggressor (a point made in the 2010 QDR), the predominant demand has been for forces that can conduct extended counterinsurgency and stabilization operations.

By the end of 2003, the demands on the Army in Afghanistan and Iraq made it clear that the Army had to sustain sizable deployments of long duration to stabilize these states. The Army leadership looked to a modular force design, coupled with a rotational readiness cycle, called ARFORGEN, to provide a steady, predictable stream of forces that could sustain such operations over an extended term.

To meet the demand, the Army began to draw on RC units on a regular, sustained basis and at a rate only somewhat lower than that of the AC. The result was predictable. A number of units began to feel overextended. While the all-volunteer force was not in danger of "breaking," it was under significant stress. This is an important consideration in the 2010 QDR, which includes "preserve and enhance the all-volunteer force" as one of four key tenets of the U.S. defense strategy.[1]

Addressing the Change in Demand

The Army took a number of steps in an attempt to configure the force in a way that optimized its organization to meet the emergent demand for land power in current and potential conflicts. Nearly concurrently with the modularity initiative, the service also embarked on the effort to grow the Army and to rebalance the force. Table 3.1 summarizes the major changes that resulted.

---

[1]   U.S. Department of Defense, *Quadrennial Defense Review Report*, February 2010b, p. v.

Table 3.1
Major Changes in Force Structure, 2002–2008

Year	Number of Units (AC/National Guard/Army Reserve)				
	Infantry Battalions	Mechanized Battalions	Cavalry Squadrons	Sustainment Brigades	Support Battalions
2002	101 (51/50/0)	55 (21/34/0)	21 (13/8/0)	28 (18/8/2)	145
2008	122 (68/54/0)	55 (38/17/0)	79 (46/33/0)	32 (14/9/9)	203

SOURCES: ARSTRUC Memo 12-17, December 15, 2009; ARSTRUC Addendum 1015, version 0902, April 28, 2009; ARSTRUC Memo 09-13, October 4, 2008; Program Objective Memorandum 08-13 ARSTRUC Message, April 7, 2006; Modular Support Forces Analysis Mini–Total Army Analysis Program Objective Memorandum 2006–2011, February 4, 2005; Program Objective Memorandum 06-11 ARSTRUC Message, April 30, 2004; Total Army Analysis 09 ARSTRUC Message, April 30, 2002; Total Army Analysis 07 ARSTRUC Message, December 22, 1999; Total Army Analysis 05 ARSTRUC Message, March 3, 1998.

The numbers in parentheses indicate the number of units in the AC, the Army National Guard, and the Army Reserve, respectively. The infantry battalion columns (infantry and mechanized battalions) include 15 infantry battalions in Stryker BCTs in 2002 and 21 in 2008. The 55 battalions reflected in the 2008 mechanized infantry column are new combined arms battalions. In addition, in 2002, there were 70 tank battalions: 33 in the AC and 37 in the National Guard. Many of the tanks were integrated in the new combined arms battalions, but according to analysis by the Congressional Research Service, 19 tank battalions left the force structure altogether.[2] The columns summarizing the change in reconnaissance, surveillance, and target acquisition (RSTA) squadrons (cavalry squadrons, sustainment brigades, and support brigades) include the squadrons in the 3rd Armored Cavalry Regiment. Sustainment brigades increased from 28 to 32 over the period, but the number of AC brigades fell from 18 to 14. The 2002 figure includes division support commands (DISCOMs) and corps support commands as brigade-equivalent formations. Individual support battalions grew in number from 145 to 203.

The number of military police (MP) battalions increased in the new force structure, as shown in Table 3.2, but the number of MP brigades fell from 13 to 11. The number of Criminal Investigation Division groups held constant at two. The number of MP battalions grew from 66 in 2002 to 78 in 2008, including four additional battalions in the AC, a reduction of five battalions in the Army National Guard, and an increase of 13 battalions in the Army Reserve.

The changing demand signal also meant that some units were less critical in the force structure. They served as "bill payers" in the move to the modular force structure, as Table 3.3 illustrates.

The number of chemical brigades fell from nine to three. Medical brigades increased from 13 to 14 overall, with the number in the AC increasing from two to four. (Corps support hospitals fell from 37 in 2002 to 26 in 2008.) Field artillery (fires) brigades dropped from 23 in 2002 to 13 in 2008.

Changes in field artillery battalions were less significant, dropping from 132 in 2002 to 123 in 2008. Moreover, more of the remaining field artillery force structure shifted into the

---

[2] Congressional Research Service, *U.S. Army's Modular Redesign: Issues for Congress*, Washington, D.C., RL32476, updated May 5, 2006.

Table 3.2
Changes in Military Police Force Structure, 2002–2008

Year	Number of Units (AC/National Guard/Army Reserve)		
	Brigades	Criminal Investigation Division Groups	Battalions
2002	13 (6/3/4)	2 (2/0/0)	66 (22/37/7)
2008	11 (5/3/3)	2 (2/0/0)	78 (26/32/20)

SOURCES: ARSTRUC Memo 12-17, December 15, 2009; ARSTRUC Addendum 1015, version 0902, April 28, 2009; ARSTRUC Memo 09-13, October 4, 2008; Program Objective Memorandum 08-13 ARSTRUC Message, April 7, 2006; Modular Support Forces Analysis Mini–Total Army Analysis Program Objective Memorandum 2006–2011, February 4, 2005; Program Objective Memorandum 06-11 ARSTRUC Message, April 30, 2004; Total Army Analysis 09 ARSTRUC Message, April 30, 2002; Total Army Analysis 07 ARSTRUC Message, December 22, 1999; Total Army Analysis 05 ARSTRUC Message, March 3, 1998.

Table 3.3
Unit Reductions, 2002–2008

Year	Number of Units (AC/National Guard/Army Reserve)				
	Chemical Brigades	Medical Brigades	Fires Brigades	Air Defense Artillery Brigades	Aviation Brigades
2002	9 (1/3/5)	13 (2/0/11)	23 (6/17/0)	6 (5/1/0)	35 (15/19/1)
2008	3 (1/1/1)	14 (4/0/10)	13 (6/7/0)	8 (6/2/0)	25 (11/12/2)

SOURCES: ARSTRUC Memo 12-17, December 15, 2009; ARSTRUC Addendum 1015, version 0902, April 28, 2009; ARSTRUC Memo 09-13, October 4, 2008; Program Objective Memorandum 08-13 ARSTRUC Message, April 7, 2006; Modular Support Forces Analysis Mini–Total Army Analysis Program Objective Memorandum 2006–2011, February 4, 2005; Program Objective Memorandum 06-11 ARSTRUC Message, April 30, 2004; Total Army Analysis 09 ARSTRUC Message, April 30, 2002; Total Army Analysis 07 ARSTRUC Message, December 22, 1999; Total Army Analysis 05 ARSTRUC Message, March 3, 1998.

AC, moving from 53 AC and 79 National Guard battalions in 2002 to 84 AC battalions and 39 National Guard battalions in 2008. Reorganization of air defense artillery increased the number of brigades from six to eight and increased the number of AC battalions by one for a total of six. These numbers obscure the fact that air defense artillery has undergone major restructuring, shedding many of its short-range accompanying air defense units and theater-centric air defense commands. Aviation brigades fell from 35 to 25, and aviation battalions from 110 in 2002 to 92 in 2008. A larger percentage remained in the AC however: approximately 44 percent of the aviation battalion force structure.

Other branches, including engineers, signal, civil affairs and psychological operations (PSYOPS), and military intelligence also underwent changes during this period, as shown in Table 3.4. The number of engineer brigades fell from 18 to 16, and the RC units were reduced from 11 to seven in the Army National Guard, while the Army Reserve grew from two to four brigades. Civil affairs and PSYOPS battalions held constant between 2004 (the earliest year for which we have data) and 2008, at 32 and 14, respectively. PSYOPS companies grew in number from 19 to 33, however. Military intelligence brigades and groups increased from six to eight, and all of these units moved to the AC by 2008. Military intelligence battalions, on the other

Table 3.4
Other Force Structure Adjustments, 2002–2008

Year	Number of Units (AC/National Guard/Army Reserve)			
	Engineer Brigades	Signal Brigades	PSYOPS Companies	Military Intelligence Units
2002	18 (5/11/2)	14 (10/3/1)	19	6 (5/0/1)
2008	16 (5/7/4)	13 (10/2/1)	33	8 (8/0/0)

SOURCES: ARSTRUC Memo 12-17, December 15, 2009; ARSTRUC Addendum 1015, version 0902, April 28, 2009; ARSTRUC Memo 09-13, October 4, 2008; Program Objective Memorandum 08-13 ARSTRUC Message, April 7, 2006; Modular Support Forces Analysis Mini–Total Army Analysis Program Objective Memorandum 2006–2011, February 4, 2005; Program Objective Memorandum 06-11 ARSTRUC Message, April 30, 2004; Total Army Analysis 09 ARSTRUC Message, April 30, 2002; Total Army Analysis 07 ARSTRUC Message, December 22, 1999; Total Army Analysis 05 ARSTRUC Message, March 3, 1998.

hand, decreased from 59 to 50 over the period, with much of the reduction occurring in the RC: 26 battalions in 2002 to 19 in 2008.

Rebalancing the Reserve and Growing the Force

In setting out to develop the modular force, TRADOC aimed to configure the force in a way that would permit the Army to carve more units out of the existing force without an increase in end strength. Additional goals were to create unit stability and to rebalance the RC so that it had the capabilities needed to support ongoing operations. The effect was to begin a transformation of the RC's focus. During the Cold War, it had served primarily as a strategic reserve appropriate for providing large formations after relatively long mobilization times.

Most recently, CSA General George W. Casey's "Grow the Army" initiative has led to the force being organized around 45 BCTs in the AC and 28 BCTs in the RC for a total force of 73 BCTs. This represents an increase from 33 to 45 BCTs in the AC and from 24 to 28 in the RC.

This total force of 73 BCTs targeted by DoD, as stated in the 2010 QDR, is composed of

- 40 infantry BCTs
- eight Stryker BCTs
- 25 heavy BCTs.

The Army's leadership typically cites 40 infantry BCTs, nine Stryker BCTs, and 24 heavy BCTs as the target force. In addition, the Army's leadership has targeted a force structure of 227 support brigades to provide the force with a full complement of CS and CSS by 2011.

The increase in the number of BCTs is enabled, in part, by the increase in end strength by 74,000 troops, which the Army requested and Congress authorized. Moreover, a greater proportion of these *additional* troops have been allocated to the MTO&E portion of the force. As a result, the MTO&E forces now make up a greater percentage of the total force than in the premodular Army. Table 3.5 shows the major changes in the Army's force structure. The figure shows the numbers of personnel positions in the AC, the National Guard, and the Army Reserve, and disaggregates these figures to reflect the strength in the operating force (OF), which represents soldiers in MTO&E units; the generating force (GF), otherwise known as the TDA (Table of Distributions and Allowances) or administrative side of the Army; and the total for the component. The lower rows in the table reflect the component's overall size as a

Table 3.5
Summary of Force Structure Changes, 2003–2011

Year	Structure	AC			Army National Guard			Army Reserve			Total Army		
		OF	GF	Total	OF	GF	Total	OF	GF	Total	OF	GF	Total
2003	Personnel	308,817	106,692	478,509	337,649	35,719	373,368	149,311	60,854	210,165	795,777	203,265	1,062,042
	% of component	65	22	—	90	10	—	71	29	—	75	19	—
	% of Army	29	10	45	32	3	35	14	6	20	75	19	—
2011	Personnel	376,404	92,086	549,790	316,753	36,992	357,745	148,850	43,734	204,584	842,007	172,812	1,112,119
	% of component	68	17	—	89	10	—	73	21	—	76	16	—
	% of Army	34	8	49	28	3	32	13	4	18	76	16	—
2003–2011	Personnel (change)	67,587	−14,606	71,281	−20,896	1,273	−15,623	−461	−17,120	−5,581	46.230	−30,453	50,077
	% (change)	22	−14	15	−6	4	−4	0	−28	−3	6	−15	5

SOURCES: FY 2003 data from the Structure and Manpower Allocation System (SAMAS) historical database and FY 2011 data as of June 24, 2010, from the SAMAS M-Force Lock.

NOTE: Trainees, Transients, Holdees, and Students and Friction accounts are not included in the table.

percentage of the Army's overall size. Several statistics stand out. First, as shown at the bottom of the table, there were reductions in the generating force across all components of the Army as manpower was pushed into the OF. The Army was able to add 67,587 soldiers to its tactical formations between 2003 and 2011. In addition, the service shrank its GF to unprecedented low levels, especially compared to the Air Force and Navy, in which the generating force typically accounts for approximately 40 percent of the total service.

Redesigning Brigades as BCTs

The preceding discussion only partially explains the approach that the Army used to form more BCTs as it developed the modular force. To make the BCTs more self-sufficient and better suited for their intended operations, the Army made some units organic to the BCT that formerly had been owned by the division. These units, most notably an RSTA squadron, artillery battalion, brigade special troops battalion, and a brigade support battalion, means the number of troops in the modular BCT is higher than in the premodular brigades. A comparison with brigades from the earlier force structure highlights the differences between the new BCTs and their predecessor formations. To perform realistic comparisons, we included the units from the division base that would have accompanied the premodular brigades as "attachments." These units included a direct support artillery battalion, forward support battalion from the DISCOM, and military intelligence and engineer support from the divisional military intelligence and engineer battalions. The MTO&E for the 2nd Brigade, 1st Armored Division, represented the "old" mechanized infantry brigade. The MTO&E for the 2nd Brigade, 3rd Infantry Division, represented the "old" tank brigade. Meanwhile, the MTO&E for 4th Brigade, 3rd Infantry Division, represented the new heavy BCT. The MTO&E for the 2nd Brigade, 10th Mountain Division, represented the "old" light infantry brigade; the 2nd Brigade, 82nd Airborne Division, represented the "old" airborne brigade; and the MTO&E for the 2nd Brigade, 101st Airborne Division represented the new infantry BCT. The premodular MTO&Es date from FYs 2003–2004, and the modular MTO&Es from FYs 2005–2006. Tables 3.6 through 3.9 summarize the points of comparison.

Table 3.6 reflects the increased numbers of logistics personnel in the new organizations relative to their premodular forebearers. The table presents a comparison of heavy brigades, contrasting the strengths of premodular tank and mechanized formations with those of the modular heavy BCT. The table does the same for infantry, contrasting the premodular light and airborne formations against the modular infantry BCT. The row labeled "Logistics personnel" reflects the number of transportation corps, ordnance, quartermaster, and multifunctional personnel assigned to the brigades. The row labeled "% logistics personnel" indicates the percentage of logistics personnel relative to total personnel. The row "Combat-to-logistics ratio" presents a ratio of combat personnel (a subset of total personnel examined in more detail in the Table 3.7) to logistics personnel. The row labeled "Tooth to tail" presents the ratio of total personnel and logistics personnel. The next row down presents figures for CS personnel: engineers, signal, MP, military intelligence, PSYOPS, civil affairs, public affairs, and chemical. The final row presents the CS strength as a percentage of the total.

Driven by the higher percentages of logistics personnel in the modular units, the combat-to-logistics and tooth-to-tail ratios are lower.

Table 3.7 offers another comparative perspective, emphasizing the units' relative ability to control space and protect populations. The table presents total personnel strength, along with the total strength (percentages) for enlisted soldiers, senior enlisted (E-7 and above), and

Table 3.6
Comparison of Support Personnel

Structure	Heavy Brigade			Infantry Brigade		
	Premodular (tank)	Premodular (mechanized)	Modular	Premodular (light infantry)	Premodular (airborne)	Modular
Total personnel	3,624	3,692	3,645	2,812	3,249	3,447
Logistics personnel[a]	652	451	731	377	450	634
% logistics personnel	18	12	20	13	14	18
Combat-to-logistics ratio	2.4/1	4.0/1	1.9/1	4.7/1	4.5/1	2.7/1
Tooth to tail (total-to-logistics ratio)	5.6/1	8.2/1	5.0/1	7.5/1	7.2/1	5.4/1
CS personnel[b]	656	593	492	348	409	413
% CS	18	16	13	12	13	12

[a] Includes transportation corps, ordnance, quartermaster, and multifunctional personnel assigned to the brigades.

[b] Includes engineers, signal, MP, military intelligence, PSYOPS, civil affairs, public affairs, and chemical.

Table 3.7
Comparative Personnel Data and Force-to-Space and Force-to-Population Ratios

Structure	Heavy Brigade			Infantry Brigade		
	Premodular (tank)	Premodular (mechanized)	Modular	Premodular (light infantry)	Premodular (airborne)	Modular
Total personnel	3,624	3,692	3,645	2,812	3,249	3,447
Enlisted (%)	92.6	92.9	91.1	92.1	92.3	90.9
Senior enlisted (%)	6.5	5.9	6.4	6.0	5.8	6.2
Officer (%)	6.9	6.7	8.0	7.4	7.1	8.2
Combat personnel, by skill (%)	43.1	48.8	37.8	63.9	62.3	50.4
Combat personnel, by role (%)	31.6	41.9	32.6	45.6	56.3	36.4
Force-to-space ratio (50 km)	22.94	30.92	23.80	25.62	36.60	25.08
Population that a brigade can patrol, by force-to-population ratio						
3:1,000	382,333	515,333	396,667	427,000	610,000	418,000
7:1,000	163,857	220,857	170,000	183,000	261,429	179,143
10:1,000	114,700	154,600	119,000	128,100	183,000	125,400
20:1,000	57,350	77,300	59,500	64,050	91,500	62,700

officers. The table also shows the percentage of combat personnel by skill identifier (e.g., MOS) and the role, representing the percentage in combat formations (e.g., rifle squads), who are available to conduct operations. The row labeled "force-to-space ratio" considers a 50-square-kilometer area of operations and shows the density of soldiers achievable based on combat personnel by role. In this regard, modular heavy BCTs have roughly the same capacity as premodular tank units but less capacity than mechanized infantry designs. The modular infantry BCT is likewise roughly comparable to the old light infantry brigade, according to this metric, but less capable than the old airborne brigade.

Finally, the last four rows reflect the unit's capacity to patrol effectively at varying densities of soldiers per 1,000 population. This metric is potentially valuable in gauging a unit's utility in counterinsurgency and stability operations, and it has historical precedence.[3] The notion is that different circumstances require different levels of control, so units that can generate larger force-to-population ratios are more suitable for situations requiring more control. The British, for example, generated 20 soldiers per 1,000 population in Northern Ireland at the height of the Troubles. Occupation in the American Zone of postwar Germany required approximately two soldiers per 1,000 population. Thus, the ability to generate appropriate force-to-population ratios can be seen as a prerequisite for satisfactory outcomes. According to this measure, the modular heavy BCT is superior to premodular tank units but significantly inferior to premodular mechanized units. The modular infantry BCT is inferior to both the premodular unit types.

The density of crew-served weapons serves as the next point of comparison between today's BCTs and their predecessor units. Table 3.8 presents the relevant data. The leftmost column in the table reflects weapon density for a 50-square-kilometer area of operations. The entries in the individual cells at the intersection of unit and weapon type indicate the density of weapons per square kilometer within that area of operations. The modular heavy BCT has superior numbers of tube-launched, optically tracked, wire-guided (TOW) antitank weapons on cavalry fighting vehicles, reflecting additional cavalry fighting vehicles assigned to the RSTA squadron and Bradley vehicles assigned to heavy units. The heavy BCT has greater sniper capabilities than either its tank or mechanized infantry predecessor and more 7.62 mm machine guns. The modular infantry BCT has more 120 mm mortars and Javelin antitank weapons than its predecessors.

The final comparative measure considers the number of vehicles in the modular and premodular brigades. The relative abundance of vehicles is important for several reasons. The first, and perhaps most obvious, reason is that more vehicles support more tactical mobility. Second, more vehicles may facilitate prompter support, because they give a unit the ability to send its own trucks to distribution points to pick up supplies. Perhaps most importantly, more vehicles means more network connectivity, because vehicles can carry and supply power to radios, mount their necessary antenna arrays, and support sensors and computers that, along with the radios, improve situational awareness. Table 3.9 presents this comparison in a way similar to that in Table 3.8; the data are based on a 50-square-kilometer area of operations and the density of vehicles within that space. The shaded cells highlight the modular BCTs' advantages according to this measure. The modular heavy BCT enjoys superiority over its premodular

---

[3] James T. Quinlivan, "Force Requirements in Stability Operations," *Parameters*, Vol. 25, No. 4, Winter 1995, pp. 59–69.

Table 3.8
Comparative Crew-Served Weapon Densities

Weapons (per km)	Heavy Brigade			Infantry Brigade		
	Premodular (tank)	Premodular (mechanized)	Modular	Premodular (light infantry)	Premodular (airborne)	Modular
Vehicle-mounted						
Tank: 120 mm main gun	1.76	0.88	1.16	0	0	0
Infantry/cavalry fighting vehicle: 25 mm chain gun	1.08	1.94	2.38	0	0	0
TOW	1.06	1.94	2.38	0.24	1.2	0.56
Artillery						
155 mm howitzer	0.36	0.36	0.32	0.36	0.36	0.32
120 mm mortar	0.24	0.24	0.28	0	0	0.24
Medium and light mortar	0	0	0	0.60	0.60	0.44
Missiles and crew weapons						
Javelin antitank weapon	1.02	0.10	0.72	1.13	1.16	1.52
.50-caliber machine gun	5.59	4.40	5.18	0.38	1.13	3.28
7.62 mm machine gun	6.24	5.92	7.50	1.49	1.52	2.78
40 mm grenade launcher	1.49	1.66	1.36	0.56	1.03	1.38
Sniper rifle	0.06	0.12	0.24	0.24	0.18	0.32

NOTE: The table shows weapon density for a 50-square-kilometer area of operations.

counterparts in all but two categories of vehicles. The modular infantry BCT is likewise superior to its predecessors, according to this measure.

Resourcing the Joint Fight

Today's force structure is the product of three initiatives: modularization of the force, the Grow the Army initiative, and the rebalancing of the force. A key benefit arising from these processes is an Army that can provide a steady and predictable supply of forces to the joint campaign. In this context *steady and predictable* has two meanings: it reflects the current force structure's ability to generate combat power and the Army's ability to provide balanced force packages. Figure 3.1 illustrates the first point.

The left side of the figure presents the old AC force structure as if it were resourcing ARFORGEN over the course of a three-year period. Several things are notable about this part of the figure. First, each annual stack of Army units is composed of many more types, and some are unique. Thus, if the 2nd Armored Cavalry Regiment were due for relief, there was

Table 3.9
Comparative Vehicle Densities

Vehicles (per km)	Heavy Brigade			Infantry Brigade		
	Premodular (tank)	Premodular (mechanized)	Modular	Premodular (light infantry)	Premodular (airborne)	Modular
Combat vehicles	3.42	4.42	4.14	0	0	0
Tracked command and control	2.74	3.06	2.26	0	0	0
Light tactical vehicles (armed)	1.00	0.88	1.32	0.30	1.34	2.08
Light tactical vehicles (other)	7.15	6.14	8.30	7.34	7.64	9.80
Medium tactical vehicles	5.07	5.10	4.68	1.61	0.35	4.72
Heavy tactical vehicles	2.41	2.12	3.26	0.16	0.10	1.28
Density, by personnel						
Combat vehicles per combat soldier	0.15	0.11	0.17	0	0	0
Tactical vehicles per soldier	0.22	0.19	0.24	0.17	0.15	0.26

NOTE: The table shows weapon density for a 50-square-kilometer area of operations.

Figure 3.1
Consistent Delivery of Combat Power

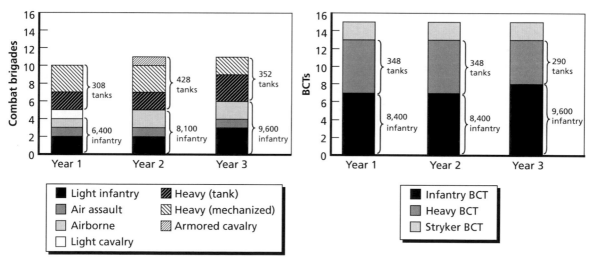

SOURCE: MTO&E data.
NOTE: Figure assumes 1:2 BOG-to-dwell ratio. BOG = boots on the ground.
RAND TR927-3.1

no other armored cavalry regiment in the force structure exactly like it; the gaining combatant command (COCOM) would have to adjust the mission and area of operations for the new unit in a way that was commensurate with its capabilities. The second notable thing about the left side of the figure is the way the amounts of infantry and armor vary over the three-year period. Given the abundance of unit types in the old force structure, it was nearly impossible to organize them in ARFORGEN to produce constant supplies of key combat capabilities.

In contrast, the right side of the figure illustrates three years' worth of ARFORGEN based on the current force structure. Each annual stack of forces draws on only three types of BCTs, and they deliver nearly constant levels of tanks and infantry. This enhanced level of consistency makes COCOM planning for Army employment easier.

Table 3.10 illustrates the current force structure's ability to deliver balanced force packages. The table indicates the AC BOG-to-dwell ratios necessary to provide support to different steady-state BCT deployment packages: a ten-BCT package, a 15-BCT package, and a 20-BCT package. The leftmost column lists supporting units that might be deployed in tandem with the BCT packages: sustainment brigades, combat aviation brigades, fires brigades, mobility enhancement brigades, and battlefield surveillance brigades. The columns to the immediate right list the number of support units available in the AC and RC, respectively. The rows beneath each BCT deployment package reflect the possible densities of support units. (Thus, 1:3 means one sustainment brigade or other support brigade for every three BCTs in the force package, and 1:6 means one such support brigade for every six BCTs in the force package.) The cells are shaded to reflect the BOG-to-dwell ratio necessary to sustain this level of support. The RC BOG-to-dwell ratio is held constant at 1:4 throughout. The lightest shading in the table indicates that that type of support can be sustained for the force package in question with an AC BOG-to-dwell ratio equal to or better that 1:2, medium shading indicates an AC BOG-to-dwell ratio that is between 1:1 and 1:2, and dark shading indicates a ratio worse than 1:1, meaning that units would be deployed more often than not.

Assuming that a division headquarters commands four BCTs, sustainment brigade levels are suitable to provide one brigade for each division and corps headquarters, with some additional flexibility. Aviation assets are suitable to deploy one combat aviation brigade with each division headquarters. Current levels for fires, mobility enhancement, and battlefield surveillance brigades are not suitable to provide assets to each division headquarters for a 20-BCT steady-state deployment while meeting BOG-to-dwell goals.

Overall Assessment

Overall, the modular force is proving to be superior to the premodular force in contributing land power to current and reasonably foreseeable joint operations. This judgment rests on four points. First, because the Army made key CS and CSS units organic to the BCTs, they are more self-sufficient than the premodular force. This means that when they deploy, they arrive with the component elements they need to prepare to move into action. The time and effort to set up and adapt to the mission demands and the environment is reduced. Moreover, the modular BCT has a larger and more capable headquarters staff (that is also organic) than did the premodular brigade. This allows the brigade to do the bulk of its own planning for complex operations of the type encountered in Afghanistan and Iraq. Such planning typically had to be done at higher-echelon headquarters in the premodular force. It also provides the BCT

Table 3.10
Providing Balanced Force Packages

Type	Component		10-BCT Steady-State Deployment					15-BCT Steady-State Deployment					20-BCT Steady-State Deployment				
	AC	RC	1/3	1/4	1/5	1/6	1/10	1/3	1/4	1/5	1/6	1/10	1/3	1/4	1/5	1/6	1/10
Sustainment brigade	13	19						1:10					1:3				
Combat aviation brigade	13	8	1:7.7					1:2.7					1:1.4	1:2.7			
Fires brigade	5	7	1:2.5					1.5:1	1:1.2	1:2.5				1.5:1	1:1.2	1:2.6	
Mobility enhancement brigade	2	19						2.5:1	1:4					2.6:1	1:4		
Battlefield surveillance brigade	3	7		1.3:1	1:3						1.3:1	1:3				1.3:1	1:3

NOTE: The table assumes RC BOG-to-dwell ratio of 1:4. The numbers of brigades for FY 2017 are based on June 2010 data from the Structure and Manpower Allocation System. Calculations were performed using an adapted version of the model developed for a RAND study of force employment options. See Lynn E. Davis, J. Michael Polich, William H. Hix, Michael D. Greenberg, Stephen Brady, and Ronald E. Sortor, *Stretched Thin: Army Forces for Sustained Operations*, Santa Monica, Calif.: RAND Corporation, MG-362-A, 2005.

AC BOG-to-dwell ratio of ≤ 1:2

AC BOG-to-dwell ratio of ≤ 1:1

AC BOG-to-dwell ratio of ≥1:1

headquarters with flexibility that its predecessors did not have for task-organizing its assets into company teams and battalion task forces; there are enough field-grade officers in the headquarters to provide leadership and staff support to whatever ad hoc formations the BCT commander concludes are necessary to accomplish the mission. It further allows the BCT to exercise an expanded span of control and more effective integration into joint operations.

Second, Army-wide, the modular force contains a greater number of BCTs than the force structure it replaced. When cycled through the ARFORGEN process the force provides a predictable supply of forces. The forces also span the spectrum—heavy, Stryker, and infantry—so a combatant commander can request the appropriate capability to match the mission. Moreover, units in the current force structure are relatively standardized, so the combatant commander will have a good idea of the capabilities of the forces in each rotation. This standardization also makes the BCTs substitutable. If one particular infantry BCT unit is not available, another with roughly the same characteristics can be called upon.

The greater number of BCTs, even with only two maneuver battalions in the heavy and infantry BCTs means that the total force has a similar number of maneuver battalions with more combat companies in total. The modular BCTs also have more organic firepower than the premodular formations. In addition, the Army can deploy a modular brigade as a coherent unit. In the premodular force, the Army could not deploy the three brigades of a division separately and effectively. If only a brigade were needed, it would be stripped from its host division, taking with it some above-brigade units. This would typically unbalance the division's capabilities so that part of the remaining force would be undeployable.

Finally, the modular BCT is better designed than the premodular brigade to participate effectively in an operation alongside with the other services. In particular, the modular BCTs' superior RSTA and communication capacity provides the foundation for improved situational awareness and close integration with other services in a joint operation.

CHAPTER FOUR

The Ability to Manage the Flexibility and Versatility of Army Forces Across the Range of Military Operations

This chapter addresses the second major issue outlined in the congressional mandate for this study, the Army's ability to manage the flexibility and versatility of its forces across the range of military operations.

Framing the Issue

DoD has a process for building force packages of "capabilities," rather than of units, to satisfy the demand for military forces. This process can unfold deliberately (for example, as a routine staff activity to provide capabilities to a contingency plan) or as a response to an emerging crisis.[1] Either way, it tends to reach into Army units to extract the specific capability sought for the task at hand and assembles these capabilities into force packages for the contingency plan or the deployment in question. In the process, divisions were tasked to provide subunits from within their MTO&E. Capabilities taken from the division base and the division support command typically left the remainder of the division unbalanced and underresourced.

The process always emphasized a "minimal footprint" which typically meant taking the minimum-essential forces to accomplish the mission. Although this line of thinking may have reached its most extreme form under Secretary of Defense Donald Rumsfeld, his insistence on lean force packages was not unique. The process itself seeks to limit the deployment of unnecessary resources. The process thus had the laudatory effect of providing custom-tailored forces for the task at hand: a battalion task force for the Sinai, an armored brigade for stability operations in the Balkans, or an aviation task force (Task Force Hawk) for combat operations against Serbia. The downside was that the parent units were often hobbled by the need to provide capabilities, and their residual forces were left less able to perform their full suite of missions.

The process of building force packages did not always proceed quickly and smoothly. Developing and deploying the force package for Operation Desert Storm took approximately six months.[2] As noted in Chapter One, the process of building and deploying Task Force Hawk took so long that the CSA at the time, General Shinseki, worried about the continued relevance of the Army if it could not respond promptly to future crises. And, if such force

---

[1] For a detailed description of the deliberate and crisis planning processes, see Joint Forces Staff College, *The Joint Staff Officer's Guide 2000*, Publication 1, Norfolk, Va., 2000, especially Chapters Four and Five.

[2] See Office of the Secretary of Defense, *Conduct of the Persian Gulf War, Final Report to Congress Pursuant to Title V of the Persian Gulf Conflict Supplemental Authorization and Personnel Benefits Act of 1991 (Public Law 102-25)*, Washington, D.C., April 1992, Chapters Three and Five.

packages included a joint task force (JTF), additional steps were involved. The JTF would need a joint staffing document to identify which staff and command positions each of the services would have to supply, a process that often required even more additional time.

Analyzing the Army's Capability to Support Joint Operations

It is against this backdrop—DoD's force packaging process—that this chapter proceeds. The core of the response is whether the current force structure does a better job of supporting force package development and deployment than its predecessor force structure (the division-centric Army) did. The answer illuminates "flexibility" (each force structure's relative responsiveness to calls for forces from the Office of the Secretary of Defense) and "versatility" (the relative abundance and variety of capabilities presented by each force structure). Speed of deployment, however, is not a part of the calculation. This exception reflects the fact that deployment speed is largely dominated by the transportation means supporting the deployment. Those speeds are constants for our purposes: somewhere between 11 and 27 days by sea from the United States, depending on the distance to the theater of operations, and less by air. This chapter focuses on the force structure's ability to respond to requests for forces.

The Army's Major Operational Themes

The Army maintains five major operational themes that inform its thinking about the tasks, conditions, and circumstances for which its forces prepare. These themes provide the context for the analysis of flexibility and versatility.[3] They are as follows: major combat operations, irregular warfare, limited interventions, peace operations, and peacetime military engagement. These operational themes allow the Army to bundle types of activities associated with each theme and thus give substance to the spectrum of military operations. We do not have empirical evidence for all the operational themes, but we consider, in the remainder of this chapter, the relative performance of the division-centric force and the modular force under circumstances for which we do have the evidence.

Performance of the Division-Centric Force Structure

The division-centric force structure changed somewhat after the end of the Cold War, growing smaller, with fewer AC and RC divisions. Several features endured over the years, however. First, until the recent rebalancing of the active and reserve components, the RC held a disproportionate share of the CS and CSS formations. This condition was by design; General Creighton Abrams believed that dependence on the RC would prevent future presidents from committing U.S. forces to combat unless they were confident that they enjoyed robust domestic support for the decision, which would manifest as support for the executive order mobilizing RC units for the operations at hand.[4] However well intended this design feature, it meant that AC units were constrained operationally by their access to support from the RC. This constraint, in turn, affected the force packaging process, especially in circumstances in which the

---

[3] See HQDA, 2008, Chapter Two. Paragraph 2-11 states, "Different themes usually demand different approaches and force packages, although some activities are common to all."

[4] Bernard D. Rostker, *I Want You! The Evolution of the All-Volunteer Force*, Santa Monica, Calif.: RAND Corporation, MG-265-RC, 2006.

need to respond promptly to a crisis could benefit from capabilities found only in the echelons above corps force, which was largely based in the RC.

In addition to the fact that the preponderance of CS and CSS resided in the RC, the AC had very limited CS and CSS capabilities. The CS and CSS in the division support command, for example, underwent fairly significant reductions as the Army moved from the Division 86 structure to that of the Army of Excellence. The maintenance battalion, the supply and transport battalion, and the medical battalion were combined to form a main support battalion. The three forward support battalions, the units that provide dedicated, primary support to the combat brigades, were also reduced in size and capability.[5] Thus, while the reorganization may have had many positive features, it reduced critical support in the AC to a minimum so that when a brigade was deployed and took its designated share of support along, it rendered the residual DISCOM units much less capable of providing full support to the remainder of the division.

This same circumstance prevailed in the rest of the Army's units. For example, if a battalion task force was formed and deployed, consequences rippled throughout its parent force structure. Its parent brigade was less capable in the battalion's absence, since the battalion typically represented one-third of its combat power and required augmentation from the brigade headquarters to function as a task force. To the extent that the battalion task force needed support from DISCOM, those demands undermined the functionality of the remainder of the command. If the task force required attachments from the division base (e.g., aviation, intelligence, engineering support), those assets were no longer available to support the nondeploying remainder of the division.

Providing headquarters for deploying force packages was not always straightforward. Corps and division headquarters had responsibilities for the entirety of their assigned forces, so it was difficult to detach them to serve a special purpose, like a contingency package headquarters that did not include all the division's subordinate units. Nor were the division and corps headquarters capable of diverse command roles; they required additional augmentation for command roles beyond command and control of their own forces. Often (e.g., Task Force Hawk), the headquarters was organized ad hoc.[6]

Performance of Today's Force Structure

Although dependencies between the AC and RC remain based on various CS and CSS capabilities, today's force structure has largely overcome the principal disadvantages associated with the division-centric structure that preceded it. First, the current AC force structure includes some of each CS and CSS unit type: aviation brigades, sustainment brigades, mobility enhancement brigades, battlefield surveillance brigades, and fires brigades. Figure 4.1 illustrates the proportions of units in each component.

Second, each BCT has a proportionately greater organic CS and CSS capability than its predecessor formations. The amount of CS and CSS is proportionately greater than even that of a task-organized BCT under the old design. The brigade special troops battalion provides command-and-control capabilities, a fire support element, an MP platoon, a signal company,

---

[5]   Edward L. Andrews, *The Army of Excellence and the Division Support Command*, Carlisle, Pa.: U.S. Army War College, May 21, 1986.

[6]   See Nardulli et al., 2002, Chapter Four.

Figure 4.1
Proportions of Headquarters and Units, by Component

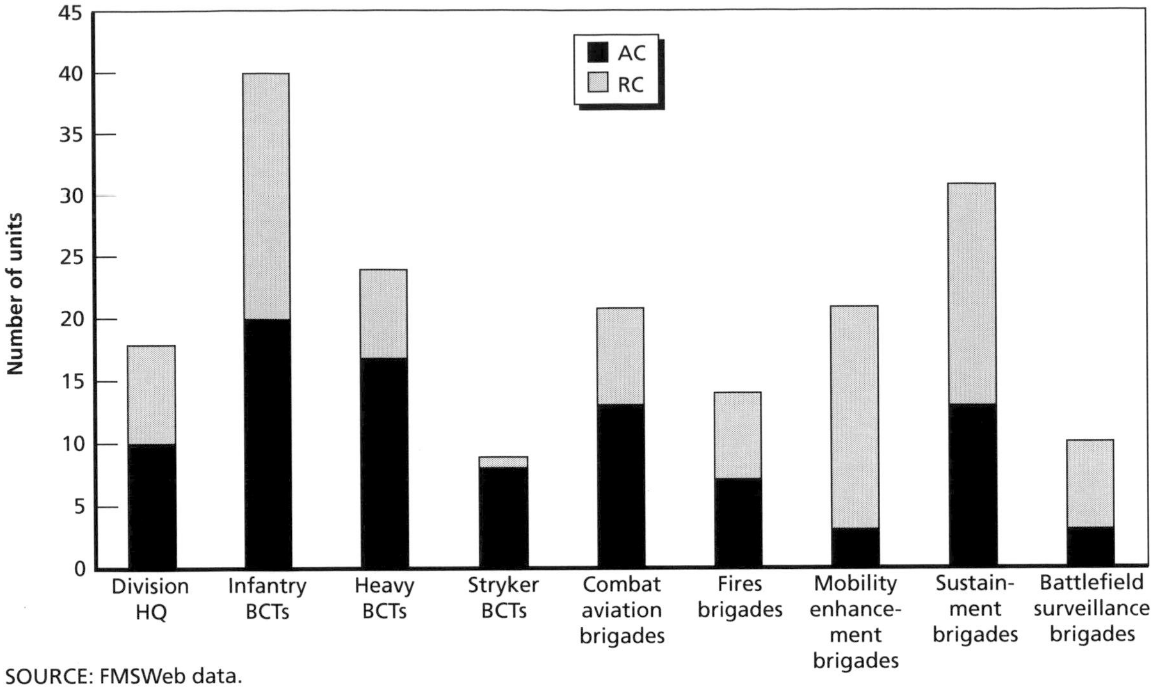

SOURCE: FMSWeb data.
RAND *TR927-4.1*

and a military intelligence company, while the support battalion includes a maintenance company, distribution company, medical company, and forward support companies.[7]

The current force structure features smaller units (BCTs and brigades rather than divisions) as its principal building blocks. Thus, force package–induced turbulence is less damaging to the force structure in terms of packaging capabilities for deployment than was the case under the earlier, division-centric force structure. For example, the effects of deploying a battalion task force to the Sinai are confined to the BCT from which it comes. Fewer units are distorted and rendered less capable (or even incapable) of performing their primary mission as a consequence of losing subordinate elements to force packaging for crises and current operations. Typically, the BCTs and brigades *are* the force packages.

Policies accompanying the current force structure have broken the habitual associations between headquarters and their subordinate BCTs or brigades. Both division and corps headquarters are employed and deployed independently. This means that finding a headquarters suitable to command a new force package is easier than under the previous force structure. The BCT and brigade headquarters have likewise gained flexibility. Current designs are capable of larger spans of control than their predecessors. Brigades typically feature "plug-ins" that anticipate the arrival of attachments (e.g., unmanned aircraft systems) to augment their capabilities.

Table 4.1 illustrates the manpower changes in the corps headquarters. It presents not only larger end strengths but also additional depth in field-grade officers, warrant officers, and senior noncommissioned officers (NCOs). These additional personnel make the current corps headquarters suitable for a variety of roles beyond corps headquarters—something their prede-

---

[7]   Information from MTO&E 87300G00, provided to the authors by HQDA G-3/5/7.

Table 4.1
Manpower Changes in the Corps Headquarters

Personnel	I Corps		III Corps		XVIII Corps	
	Premodular	Modular	Premodular	Modular	Premodular	Modular
Enlisted	207	485	207	531	205	502
Senior enlisted (E-7 or higher)	43	154	43	154	45	153
Officer	119	241	119	243	119	242
Company grade	17	45	17	53	17	46
Field grade	99	192	99	186	99	192
General grade	3	4	3	4	3	4
Warrant	1	51	1	51	1	51
Total	327	777	327	825	325	795

SOURCE: Data from FMSWeb, as of May 2010.

cessor organizations could not have done without augmentation. Now, they can also serve as JTF, Army forces, or combined forces land component command headquarters.

Table 4.2 presents the changes in division headquarters resulting from the current force structure. As with the changes to corps headquarters, the additional manpower and capabilities residing in the current division headquarters make them capable of larger spans of control and equip them for a wider suite of command-and-control responsibilities, including task force and JTF.[8]

The current force structure benefits from another attribute that was less prominent in the division-centric force structure: the ability to replicate units among its forces. As Figure 3.1 in Chapter Three illustrated, the uniformity within types of units and the relative abundance of types of units means that it is easier for the current force structure than for its predecessor to provide a steady supply of well-understood capabilities spanning the spectrum of light-through-heavy forces.

Assessing Flexibility and Responsiveness

The responsiveness of a force structure to demands arising from force packaging is a good measure of flexibility, and the abundance of different, replicable capabilities available within a given force structure is a good measure of flexibility. This leads us to conclude that the current force structure exhibits greater flexibility and versatility than the force structure it replaced. The current force structure's superior responsiveness lies in its ability to promptly provide units as building blocks for force packages without the disruptions and consequences that characterized the efforts of the earlier force structure. The current force structure provides easily accessible building-block units and a variety of headquarters suitable for the full range of command circumstances with less need for augmentation than their predecessors. The current force

[8] This is true under some conditions. Employment as a JTF headquarters would require augmentation from the other services involved.

Table 4.2
Changes in Division Headquarters

Personnel	4th Infantry Division		82nd Airborne Battalion		29th Infantry Division		36th Infantry Division (Army National Guard)	
	Premodular	Modular	Premodular	Modular	Premodular	Modular	Premodular	Modular
Enlisted	217	555	165	602	158	590	177	594
Senior enlisted (E-7 or higher)	67	140	33	123	34	122	34	122
Officer	102	198	72	209	76	208	78	208
Company grade	33	68	27	78	28	78	34	78
Field grade	66	127	42	128	45	127	41	127
General grade	3	3	3	3	3	3	3	3
Warrant	22	45	4	41	4	40	4	41
Total	341	798	241	852	238	838	259	843

SOURCE: Data from FMSWeb, as of May 2010.

structure's responsiveness is also superior because of a better mix of its organic capabilities—combat, CS, and CSS—than was the case in the earlier force structure.

The current force structure features superior versatility relative to the division-centric structure. This superior versatility is a result of the fact that, as discussed in Chapter Three, the BCTs are generally better armed and staffed than the units they superceded. Furthermore, the full range and quantity of CS and CSS capabilities is greater in the AC force structure than it was in the division-based force structure, the abundance of capabilities from which to choose within the AC is richer than in the division-based force structure, and the number of BCTs that the ARFORGEN process can produce is greater. These advantages, in our judgment, outweigh the disadvantage of two rather than three maneuver battalions.

CHAPTER FIVE

Risk Associated with the Current Force Structure

This chapter examines the third major congressional question concerning the tactical, operational, and strategic risks associated with the heavy, medium, and light modular combat brigades and functional support and sustainment brigades.

Framing the Issue

Risk is a critical consideration in military force planning. It cannot be eliminated entirely, though it can be mitigated through careful planning. That planning begins with a systematic review of the missions that the military is likely to be called on to execute and a judgment of how well the forces would do across the spectrum of these likely missions. If we understand risk as the product of probability and consequences ($R = P \times C$), then risk can be mitigated by either reducing the probability of the key events or reducing their attendant consequences. This calculus captures all levels of military operations: strategic, operational, and tactical.

The Problem

The 2010 QDR defined strategic and operational risk and the elements that contribute to risk.

Strategic risk captures the ability, or shortfall in the ability, to execute defense priority objectives in the near term, midterm, and long term in support of national security. The military dimension of this risk is the ability of U.S. forces to adequately resource, execute, and sustain military operations in the near- to midterm and in the mid- to longer term.

Operational risk captures the ability, or shortfall in the ability, of the force to execute strategy successfully and within acceptable human, material, financial, and strategic cost parameters. Consideration of operational risk requires assessing DoD's ability to execute current, planned, and contingency operations in the near term.

From these higher-level risks, it is possible to deduce a working definition of tactical risk. *Tactical risk* includes a unit's ability to accomplish its immediate mission (e.g., seize and hold terrain, secure a population) in support of operational goals as stated by the commander.

Manifestations of Risk

For Army units, the various risks manifest in two principal forms: (1) as more capable opponents and (2) as insurgencies, stability operations, and irregular wars whose scale exceeds the Army's resources to prevail. (That is, the Army cannot generate appropriate force-to-space or force-to-population ratios to provide stability and security.) These considerations form the context of our evaluation of risk, outlined in the next section.

Evaluating the Risk

To evaluate risk, we compared the current force structure's ability to mitigate risks against the division-centric, premodular force structure's ability to do so. The research team measured this ability in terms of reducing the probability associated with the future value calculation of risk (R = $\boldsymbol{P}$ × C). The hypothesis at work in the analysis is that a larger, more capable force comprising more deployable subordinate units can do a better job of driving $\boldsymbol{P}$ downward than a smaller, less capable force with fewer and less capable subordinate units. We drew on the numerical analysis of MTO&Es and overall force structure attributes presented earlier in this report to support the comparison of modular force and division-centric force risk mitigation.

We compared modular units (with a focus on BCTs) with their premodular predecessors to determine whether there were any clear differences in the risks they could mitigate. We conducted a survey of former BCT commanders, specifically to understand their views on the risks associated with the current and earlier force structures. We sought former BCT commanders whose battalion command tours had been in the premodular force in the belief that command experience in both force structures would give them the necessary perspective on the strengths and shortcomings associated with each. We augmented their perspectives with those found in two key documents in which experts compared their experiences with the modular force and premodular formations. These documents were *Former Brigade Commander Seminar #2* (May 2009) and TRADOC Army Capabilities Integration Center's *ARCIC Comprehensive Lessons Learned* (August 2009).[1]

Former BCT Commander Interviews

We completed semistructured interviews with former BCT commanders who deployed to Iraq. Their BCTs operated in a variety of situations: Some conducted operations in dense urban environments in Baghdad while others were deployed across large areas of operations that encompassed several key cities. Most reported that total risk declined under the new BCT design, primarily because of the new resources available organically within the brigade and the corresponding increase in responsiveness and flexibility.

One former commander indicated, "Risk went down. The modular BCT trained and then deployed as a team. The increased capabilities of the enablers allowed the commanders on the ground to reduce risk to our soldiers and to the mission." The consensus that emerged from the interviews was that "modular transition mitigated risks."

Analysis of Strategic Risk

Beginning with *strategic risk*, the QDR provides a projection of the capabilities it will be looking to the Army to provide. The most prominent theme captured in the "Defense Strategy" section of the 2010 QDR is the imperative to support the needs of combatant commanders and, through them, the forces in the field today. Simply put, the imperative is to *prevail in today's*

---

[1]   Brian Layton, Office of the Deputy Chief of Staff G-8, U.S. Army, *Former Brigade Commander Seminar #2*, May 13, 2009; U.S. Army Training and Doctrine Command, *ARCIC Comprehensive Lessons Learned*, August 2009.

wars. Naturally, this is focused principally on U.S. operations in Afghanistan and, to a lesser degree, Iraq. This imperative is made explicit and stated right at the outset of the report: "The first [objective] was to further rebalance the capabilities of the U.S. Armed Forces and institutionalize successful wartime innovations to better enable success in today's wars while ensuring that our forces are prepared for a complex future."[2] In other words, DoD acknowledges the progress of the military—the Army, in particular—in organizing and equipping for "today's wars" while quite clearly saying that the services must push further.

The QDR goes on to reinforce this focus, stating that operations of the type in Afghanistan and Iraq are not one-shot deals. "The character of these wars—with enemies hiding among populations, manipulating the information environment and employing a challenging mix of tactics and technology will be an important part of the future spectrum of conflict."[3] In this context, another important tenet of the "Defense Strategy" section of the QDR is the imperative to preserve and enhance the all-volunteer force.

In the mid- to long term, the QDR goes on to say, DoD will "expect there to be enduring operational requirements in Afghanistan and elsewhere to defeat Al Qaeda and its allies."[4]

The Army, in transitioning to the modular force design, has taken a step in the direction of addressing this strategic imperative. The modular design, with the support of the ARFORGEN process, can provide combatant commanders with a steady, reliable stream of available forces with well-understood capabilities to execute operations typical of *today's wars*. Figure 5.1 provides an overview of trends in supply and demand. The figure illustrates the anticipated drawdown in demand overlayed with the numbers of soldiers available to satisfy remaining demand at differing ratios of time deployed to time out of the combat theater, typically referred to as BOG-to-dwell ratios. In each set of bars, the leftmost bar represents sufficiency based on AC soldiers alone. The rightmost bar reflects sufficiency based on the AC augmented by 25,000 RC soldiers.

By organizing a total of 73 BCTs with supporting structure, the modular force will have a reservoir adequate to cope with today's wars and operations that could reasonably arise in the future, as Figure 5.1 suggests.

The drop in demand for deployed forces as U.S. Army deployments in Iraq wind down and as deployments to Afghanistan level off and then gradually decline should allow the Army to adjust the BOG-to-dwell ratio for its forces to 1:2 in the AC and 1:4 in the RC, its stated interim target. With the programmed target of 45 BCTs in the AC and 28 BCTs in the RC, this would provide a pool of roughly 15 available BCTs in the AC and five in the RC per ARFORGEN cycle.

These force levels are more than adequate to cope with the ongoing intended deployment of forces to Afghanistan until the draw-down. As U.S. troops are steadily drawn down there, the Army could adjust to a BOG-to-dwell ratio of 1:3 for the AC and 1:5 for the RC, the Army's stated eventual target. The Army regards these BOG-to-dwell ratios as sustainable over the long term—that is, ratios that will preserve the health of the Army's all-volunteer force. This would support 11 available BCTs in the AC and four in the RC per ARFORGEN cycle, providing for one stabilization operation of a mid-to-large size.

---

[2] DoD, 2010b, p. 1.

[3] DoD, 2010b, p. 1.

[4] DoD, 2010b, p. 44.

Figure 5.1
Trends in the Supply of and Demand for Army Forces

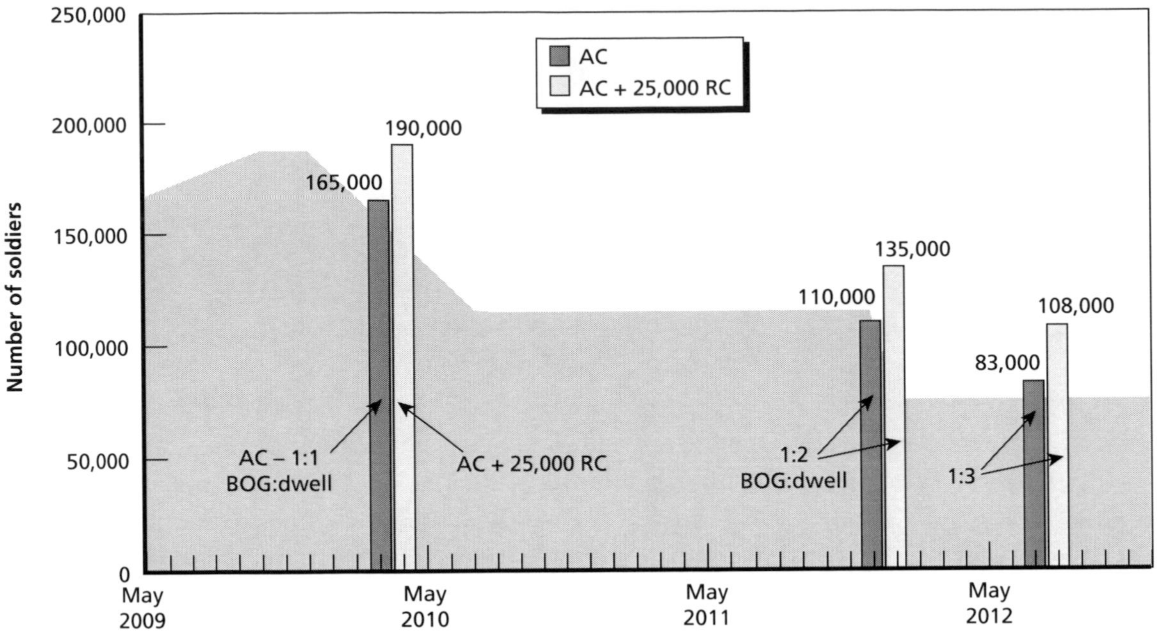

SOURCE: Derived from data in Timothy M. Bonds, Dave Baiocchi, and Lauri L. McDonald, *Army Deployments to OIF and OEF*, Santa Monica, Calif.: RAND Corporation, DB-587-A, 2010.
RAND *TR927-5.1*

This supply of forces would have been much more difficult to achieve with the premodular force that was organized into 33 brigades in the AC and 19 in the RC. The BOG-to-dwell ratio of 1:3 for the AC and 1:5 for the RC, which the Army felt was tolerable over the long term while preserving the health of the all-volunteer force, would yield only eight BCTs from the AC and three BCTs from the RC, leaving only enough for one midsized stabilization operation.

Analysis of Operational Risk

Operational risk is also mitigated by the modular force. In the irregular warfare threat that typifies today's wars, adaptability is critical. The presence of key units organic to the BCT—most notably, the RSTA squadron and a more robust brigade headquarters—provide it with an ability to plan and execute more complex operations and to gain a better situational awareness of the battlespace. Brigades with specialized capabilities (e.g., aviation, fires, sustainment, mobility enhancement, battlefield surveillance) can be deployed with the BCT to tailor it for the situation it will face in theater.

Although concerns have been expressed about the deficiencies that accompany the infantry BCTs and heavy BCTs, which have only two rather than three maneuver battalions, former BCT commanders did not believe that this design attribute contributed to risk. When asked explicitly, most rejected trading enablers, like the RSTA squadron, for a third maneuver battalion. In addition, the Army has been aware of the limitations that attend the design and has identified compensatory actions.[5]

---

[5] Headquarters, U.S. Department of the Army, "Answers to CSA Questions," task force modularity briefing, August 20, 2004.

One potential source of operational risk lies with more capable adversaries, though none looms on the horizon. Today's force structure has been developed in response to the adversaries at hand. Therefore, it arguably faces less operational risk than its predecessor force structure when current adversaries are considered. If more capable foes begin to emerge, the Army would have time to adapt accordingly. It has done so a number of times in the recent past: the modern battlefield studies after Vietnam to refocus on the Soviet threat; the AirLand Battle concept to leverage Air Force capabilities along with its own and to fight outnumbered and win; and Force XXI and the Army After Next in contemplation of future fights.

Other potential sources of operational risk are insurgencies, stability operations, and irregular wars whose scale is beyond the scope of Army resources. This is not an unfamiliar problem for Army commanders, some of whom have commented on the difficulties of generating force-to-space ratios suitable for current operations in Afghanistan and Iraq.[6] That said, the question is whether the current force structure does a better job of coping than its predecessor. Considered in this light, the current force structure does a better job at managing operational risk because its MTO&E strength is a larger percentage of overall end strength than in the predecessor force (see Table 3.5 in Chapter Three). In addition, the organization of RSTA and military intelligence in today's force structure, and the ongoing modernization and incorporation of technology (principally, communication and situational awareness systems) from the Future Combat System program render the current force structure superior to its predecessor in terms of capabilities for counterinsurgency, stability operations, and irregular warfare.

Analysis of Tactical Risk

Concerning tactical risk, on balance, it is hard to form a clear preference. On the one hand, the Brigade Commanders' Seminar and ARCIC Lessons Learned documents captured a desire on the part of the respondents for a third maneuver battalion in the infantry and heavy BCTs (the Stryker BCTs have maintained the three-maneuver-battalion structure). On the other hand, most respondents reported that risk went down overall under the modular design due to the additional capabilities, organic to the BCT, that improved the ability to mitigate risk.

The lack of the third maneuver battalion would limit the modular BCT's ability to generate effective force-to-space ratios, though our own analysis of the modular force structure shows that it is only one maneuver company short of a three-maneuver-battalion organization. There are only modest differences in force-to-space ratios. There are more crew-served weapons in the modular BCT, and its superior communication and RSTA capabilities were cited as critical enablers by many of the brigade commanders. Finally, there were no reports of a BCT being forced to cede ground to an enemy attack or being unable to assist a heavily engaged subordinate unit because of the lack of the third maneuver battalion.

Another criticism encountered during our interviews with former BCT commanders was that the RSTA squadron was "unable to fight for information." However, based on a comparison of their MTO&Es, RSTA squadrons have superior armament compared with premodular infantry brigade reconnaissance elements, as well as more vehicles. We suspect that this criticism reflects the perspective of a heavy BCT commander, to whom the RSTA squadron looks meager compared to a combined arms battalion—replete with tanks and infantry fighting vehicles—or a premodular cavalry squadron, like those previously found in armored cavalry

---

[6] Brian G. Watson, *Reshaping the Expeditionary Army to Win Decisively: The Case for Greater Stabilization Capacity in the Modular Force*, Carlisle, Pa.: U.S. Army War College, August 2005.

regiments. We conclude that the RSTA squadron's capabilities are not a source of increased tactical risk.

Finally, a larger portion of the end strength has been moved into MTO&E units, where it is deployable. The additional field-grade officers in the modular BCTs provide more flexibility in organizing and leading task-organized subunits. These steps collectively reduce tactical risk and better suit today's units to mitigate it.

In sum, we concluded that the modular force, relative to the premodular force, reduces strategic and operational risk and does not entail additional tactical risk.

CHAPTER SIX

The Required and Planned End Strength of the Army

This chapter addresses the final major congressional issue, the required and planned end strength of the Army and whether the move to the modular force structure required additional end-strength growth beyond that generated by operations under way at the time.

Framing the Issue

Required and planned Army end strength are important in the story of the move to the modular force structure in several respects. First, there was the expectation that, absent other factors, the Army could make the transition to the new force structure within its extant end strength. Second, as the war in Iraq unfolded into a persistent counterinsurgency and stability operation, a consensus emerged among the Army's leadership that an end-strength increase would be needed to manage pressure from repeated deployments and to facilitate resourcing the war with the essential MOSs.

The Problem

Could the transition to the modular force structure have been made within the parameters of the Army's end strength at the time? Was the density of critical MOSs in the force structure sufficient to staff the modular force? Or did modularity make its own, independent demands for additional end strength? These questions are addressed in the remainder of this chapter.

The Army Response to Increased Demand for Forces

As the Army experienced successive rotations in Iraq and Afghanistan, the transition to the current force structure intersected with other Army efforts to reduce stress on its soldiers, especially as the demand increased for greater numbers of troops in theater. The Army's response to mitigate the stress has been as follows:

1. Increase its active operating force strength within the existing end strength to 355,000.
2. Secure an increase in end strength.
3. Seek more combat effectiveness from the end strength.
4. Develop a force configuration that is compatible with the ARFORGEN process to ensure that forces "on the books" will be available on schedule.
5. Avoid putting undue stress on the soldiers and families of the all-volunteer force.

Increasing the Active Operating Strength

In 2006, the Army undertook steps to reduce the size of its TDA force and noncombat supporting force structure.[1] This step involved "converting other formations into modular . . . units . . . [and] into tailorable and strategically responsive force packages."[2] Although there was some growth in the number of logistics battalions, the Army disestablished nine field artillery battalions, 13 air defense artillery battalions, 19 engineer battalions, and 15 armored battalions to free manpower for the transition to the new, modular force.[3] The U.S. Government Accountability Office expressed misgivings in a report issued in 2005, warning that the AC end strength of 482,400 might be inadequate to staff the modular force. In particular, the authors cautioned that there were "significant shortfalls in the Army's capacity to . . . staff units."[4] The report noted there might be a shortfall of some 9,000 military intelligence specialists.

Securing an Increase in End Strength

In FY 2007, the Army sought and secured authorization to increase its end strength by 65,000 troops in the AC and 9,000 in the RC to reach a total of 547,000 in the AC, 205,000 in the Army Reserve, and 358,000 in the Army National Guard. In 2009, it secured a temporary increase of an additional 22,000 troops in the AC for an end strength of 569,000 by the end of FY 2012. This has permitted the Army to formulate its plan for the combat and support forces that will underpin a modular 73-BCT force.

Moving from the premodular to the modular force ultimately resulted in an increased requirement for troops. Most of this increase was due to the Army's decision to form more modular BCTs than the number of premodular combat brigades. Additional manpower demands have come about because units that used to be held at the division echelon, and meted out to subordinate brigades as needed, have become organic to each brigade. The efficiency inherent in holding forces in a pool and allocating them as needed has thereby been lost for the units that have been made organic to the BCT.

Seeking More Combat Effectiveness Within a Given End Strength

The Army leadership has judged that, at a given end strength, it can tailor its forces in a way that allows them to be employed more efficiently. More specifically, it has sought to organize units in a way that is appropriate to the task of extended stabilization and counterinsurgency operations—"today's wars." This led the Army's leadership to make a number of units that used to be held at the division level organic to the BCT (e.g., an RSTA battalion).

To compensate for the need for more troops to establish these organic units and to create more BCTs, the 40 infantry BCTs and the 24 heavy BCTs are being formed with two maneu-

---

[1] HQDA response (dated June 2, 2006) to U.S. Government Accountability Office, *Army Needs to Provide DoD and Congress More Visibility Regarding Modular Force Capabilities and Implementation Plans*, Washington, D.C., GAO-06-745, September 2006.

[2] HQDA G-3, "BCT Hill Response v7b," dated April 26, 2006.

[3] ARSTRUC files provided by HQDA G-3, various years.

[4] Sharon Pickup and Janet St. Laurent, Defense Capabilities and Management, U.S. Government Accountability Office, *Force Structure: Preliminary Observations on Army Plans to Implement and Fund Modular Forces*, testimony before the Subcommittee on Tactical Air and Land Forces, House Armed Services Committee, Washington, D.C., GAO-05-443T, March 16, 2005, p. 5.

ver battalions instead of the three that had been the norm in the premodular force. The nine Stryker BCTs continue to have three maneuver battalions.

This way of organizing does create a force that is more responsive to today's wars. It can be deployed in smaller packets and therefore requires less time to mobilize, be transported, and organize once in theater. The result is that the modular force can respond more quickly than could the premodular force.

If the threat environment were to change and, say, a high-end competitor were to emerge, the premodular organization could provide the more applicable combat power. This is particularly true for traditional conventional operations on a linear battlefield. Here, the ability of the echelons above brigade to husband key capabilities and mete them out to subordinate units as need demands would be a more efficient use of available manpower.

Configuring the Structure to Be Compatible with the ARFORGEN Process

The modular force configuration ensures that, within the end strength, there are enough BCTs to rotate through the ARFORGEN cycle to maintain a steady and predictable flow of ready units.

This is a strong step in the direction of preserving the health of the all-volunteer force, since troops can predict and plan for their deployments and the frequency of stop-loss orders can be greatly reduced.

The modular force organization has moved a greater portion of end strength into the MTO&E force. In 2003, 309,000 troops (or 65 percent of the total end strength) were in the MTO&E of the AC. With the increase in end strength, along with conversion to the modular design, the MTO&E is on track to rise to 376,000 troops, or 68 percent of the AC, in 2011.

Other Issues Affecting Staffing Capabilities

Absent the demands posed by two ongoing conflicts, end strength might have been sufficient to staff the modular force, but other issues influenced the actual outcome. For example, despite the Army's shift of manpower to MTO&E units and the shrinking of its TDA strength to unprecedented low levels, the move, though efficient, may have consequences. For example, the Congressional Budget Office noted that the resulting pressure on the institutional Army might pose risk to its strategic planning capabilities.[5]

So, how should the Army's stress management strategy be understood: as a success or as a failure? It appears on track to accomplish, in broad outline, what it set out to do. It is providing resources to current operations, it has increased the deployable percentage of its force structure, and it has created a modular mix-and-match-as-required force structure, whose pieces can be fitted together as circumstances warrant, with minimal disruptions to other units in the force structure. It has also preserved the all-volunteer nature of the Army. To be sure, preserving the all-volunteer force was not solely the result of the modularity and grow-the-force initiatives. Enhanced bonuses and enriched pay scales also played a role. Nevertheless, the steps taken by the Army with regard to end strength had their intended effects.

---

[5] Congressional Budget Office, *An Analysis of the Army's Transformation Programs and Possible Alternatives*, Washington, D.C., No. 3193, June 2009.

Other Factors to Consider

Section 344 of the Defense Authorization Act also directed the consideration of specific factors as part of this study:

- the Army's historical experience with separate brigade structures
- the original Army analysis or other relevant analyses, including explicit or implicit assumptions, upon which the modular BCT, functional support and sustainment brigades, and higher headquarters designs were based
- subsequent analyses that confirmed or modified the original designs
- lessons learned from OEF and OIF, including identification and analysis of how modular brigades or formations were task-organized and employed that may have differed from the original modular concept and how that confirmed or modified the original designs
- improvements that the Army has made or is implementing in brigade and headquarters designs
- the deployability, employability, and sustainability of modular formations compared with that of the corresponding premodular designs of such formations.

This appendix addresses each of these factors in turn.

The Army's Experience with Separate Brigades

The term *separate brigade* has a specific meaning within the framework of Army organizations.[1] Separate brigades historically have been independent—that is, not subordinate to a division. Unlike divisional brigades, which could expect their parent division to augment their basic capabilities with other divisional assets (e.g., engineers, aviation) through the process of task organization, separate brigades were typically designed to perform a specific function, and most of the requisite capabilities were built into them. Historically, separate brigades have played at least five distinct roles in the U.S. Army.

Separate Brigades as Building Blocks for Army Growth
Separate brigades have served as the basic components of the Army's growth. For example, as part of the buildup to its involvement in Vietnam, the Army activated three new separate brigades: the 196th Infantry Brigade, constituted in September 1965; the 199th Infantry Brigade,

---

[1] This discussion draws heavily on John B. Wilson, *Maneuver and Firepower: The Evolution of Divisions and Separate Brigades*, Washington, D.C.: U.S. Army Center of Military History, 1998.

activated in June 1966; and the 11th Infantry Brigade, activated the following month. As part of the same initiative, the Army created separate infantry brigades in the RC to bolster the strategic reserve.

Brigades as Purpose-Built Formations

Separate brigades were usually activated and designed with a specific purpose in mind. The 199th Infantry Brigade was originally intended to protect the Long Binh/Saigon area, where U.S. forces had a large, vulnerable logistics complex. The 198th Infantry Brigade was originally conceived as a blocking force along the demilitarized zone that separated the Republic of Vietnam (South Vietnam) from its northern counterpart, the Democratic Republic of Vietnam (North Vietnam). In fact, other forces were employed in this role before the 198th arrived in country, so the brigade undertook other operations.

The 193rd Infantry Brigade was created to defend the Panama Canal Zone when the canal was U.S. property. Given its mission, the 193rd was originally organized as mechanized infantry. Later, after the United States relinquished ownership of the canal, there was no requirement to defend it, so the brigade's structure was adapted for its new, more limited security role, and it became a light infantry formation.

The 173rd Airborne Brigade was originally created as a two-battalion, rapid-reaction force based on Guam. Its structure emphasized ease of transport over combat power. Later, however, when the brigade was deployed to Vietnam, an infantry battalion was added, making it more capable.

The Berlin Brigade also enjoyed a structure unique to its role in the American sector of Berlin. It was the only separate brigade in the force structure with an organic tank company. It also featured improved communication capabilities to link it directly to the U.S. Army Europe and U.S. European Command command-and-control systems.

Finally, there were the separate brigades at Fort Benning and Fort Knox, the 197th Infantry Brigade and the 194th Armored Brigade, respectively, which supported the infantry and armor schools by furnishing troops and small units for field problems and exercises that contributed to the training of the officer and enlisted student populations. As so-called school troop requirement brigades, their organizations were shorn of CS and CSS capabilities because they were not expected to be deployed; they simply supported training.[2]

Brigades as Experimental Units

Infrequently, brigades have been employed as experimental or test units. During the 1970s, the Army renewed its interest in deep attack aviation. On February 21, 1975, it activated the 6th Cavalry Brigade (Air Combat). This brigade was the first all-aviation cavalry unit (earlier formations had included ground force components), and it was conceived as a corps-level deep attack asset, able to strike into the depth of Soviet territory to inflict fatal damage on its forces.

Brigades as Instruments of Force Integration

Beginning in 1973, as it refocused its attention on Europe and the Soviet Union, the Army concluded that its many National Guard infantry brigades were not suitable as reinforcements in the event of a war in Europe. In response, the Army began reconfiguring them as mecha-

---

[2]   During the 1970s, these units were integrated into the strategic reserve but never deployed.

nized infantry and armored units: heavier formations more suitable to the threat looming along the inner-German boundary.[3]

Subsequently, brigades became the unit for integrating National Guard and AC forces. Under the "round-out" concept, AC divisions were structured with two active brigades and a third, National Guard, round-out brigade that would join the division in the event of an emergency that resulted in an executive order to federalize the National Guard brigades.

Historical Significance of Separate Brigades

As the preceding discussion suggests, separate brigades have been important to the Army for several reasons. First, their size seems to have made them suitable as a building block for the larger force structure. Activating new brigades was administratively and logistically less difficult than activating an entire division. Second, they have been efficient ways to create new capabilities to address specific tactical problems (e.g., defend the Panama Canal). Third, they have proven to be useful test beds for new capabilities (e.g., air cavalry). Finally, they have been building blocks for the integration of the AC and RC.

Analyses on Which the Modular Brigade Combat Team, Functional Support and Sustainment Brigades, and Higher Headquarters Designs Were Based

As noted in Chapter Two, modularity is a concept that has been around for nearly 20 years, beginning in the CSS realm and then appearing as an attribute of the Army After Next "strike force." When General Schoomaker became CSA on August 1, 2003, he took up the matter of the Army's force structure almost immediately, embracing modularity in the process. On September 5, 2003, he issued his initial guidance. The proposed force structure should be as capable as current units, be more deployable, and create more combat force structure than the division-centric force of the day.[4]

Concept and design development began in October, and, by February 2004, the CSA was approving designs for "modular 3 and 2 star headquarters" suitable for "assignment to regional combatant commanders to command and control Army, Joint, and multinational forces."[5] Over the course of the month, the General Schoomaker approved designs for an armored brigade unit of action (UA), an infantry brigade UA, a support UA, and two headquarters, a unit of employment X (corps/division), and a unit of employment Y (Army Service Component Command/corps).[6] The Army's TRADOC Analysis Center conducted the bulk of the analysis of the new unit designs, including a feasibility analysis. HQDA, G-8, provided the cost analysis for the new force structure's manpower and equipment requirements.

---

[3] Strictly speaking, these were not separate brigades but divisional brigades that plugged into U.S. Army Europe force structure.

[4] HQDA internal memoranda. At the time that General Schoomaker issued his guidance, there were 326,570 soldiers serving overseas in 120 countries, according to the Army's 2004 posture statement (see U.S. Army, *2004 Posture Statement*, February 5, 2004, p. 15).

[5] U.S. Army Training and Doctrine Command, Army Capabilities Integration Center, "Overarching Modular Briefing to CSA Backups v 1.2 011500," February 2004.

[6] ARCIC, 2004.

The Army engaged the regional COCOMs to ensure that the new force structure would satisfy their requirements. It also collaborated with U.S. Joint Forces Command to ensure that the emerging force structure would be compatible with the other services across seven critical functions, including operational planning, airspace command and control, fire support planning and targeting, attacking, logistics, collection and dissemination of intelligence, and the ability to create and update a common operating picture.[7]

The Army also consulted outside experts on its force structure deliberations. It hosted a "devils advocate/graybeard" meeting on November 20, 2003, and an "experts' day" on January 6, 2004, to seek the reactions and insights of retired senior officers and other subject-matter experts with regard to the evolving force designs. Insofar as internal Army briefings and memos capture their feedback, the experts raised few issues with the battlefield performance of the proposed unit designs.[8]

A Force for Today's Fight

The analysis seems to have focused on creating a force that could respond to the ongoing operations in which the Army was engaged. Although the prospective modular force would have a rich set of capabilities—indeed, a richer suite of heavy, motorized, light, and special operations forces than the force structure it would eventually supplant—these capabilities were organized and packaged for today's adversaries and the military problems they pose. Thus, the force designs would be applied during "reset" as units returned from their deployments. They would be reorganized "into modular designs that are more responsive to regional Combatant Commanders' needs; that better employ joint capabilities; that reduce deployment time; and that fight as self-contained units in non-linear, non-contiguous battlespaces."[9]

Subsequent Analyses That Confirmed or Modified the Original Designs

There is flexibility inherent in the force structure that resulted from General Schoomaker's initiatives. Although it packaged forces in such a way as to make them suitable to engage with adversaries that the Army was encountering in Afghanistan and Iraq, the structure's inherent flexibility meant that if an adversary with different attributes were to emerge, the Army could repackage its capabilities to counter the new foe. Operations in Iraq and Afghanistan also prompted some new coping mechanisms that allowed deployed units to provide feedback to the Army about how well their units performed and what modifications were required. The Joint Urgent Operational Needs Statements and Operational Needs Statements processes forwarded feedback from the field to HQDA, while the rapid equipping and Capabilities Development for Rapid Transition initiatives were part of the rapid response to these emerging requirements.

In addition to creating rapid-response mechanisms to deal with feedback, the Army also undertook an analytical review of its force structure in 2007. It created an Army Evaluation Task Force for the job. On February 7, 2007, the Army released a report to Congress, *Army*

---

[7]   ARCIC, 2004.

[8]   ARCIC, 2004.

[9]   U.S. Army, 2004, p. 16.

Transformation: Report to the Congress of the United States, February 2007. In it, the Army reported on lessons learned, "force management reviews," and similar activities that it had undertaken that confirmed or modified the designs of the modular force.[10]

Lessons Learned About Modular Brigade Task Organization in Operations Enduring Freedom and Iraqi Freedom

Congress also requested additional information about lessons learned from OEF and OIF, including how modular brigades were task-organized and employed and the ways in which these outcomes differed from the original modular concept, as well as whether the original design had been confirmed or modified. We address this package of factors by unpacking them into discrete issues. This section first treats lessons learned from OEF and OIF, then addresses the issue of task organization.

Lessons Learned from Operations Enduring Freedom and Iraqi Freedom

The Army indicates that it has learned many lessons from these operations, and that many of these lessons have led to the acquisition of new or additional equipment or the development of accelerated processes.[11] Insofar as accelerated processes are concerned, the Army indicates that it is providing effective, affordable equipment now (that is, getting the best available into the hands of its troops as quickly as possible). It is also reducing the time needed to field new and updated material solutions. Time savings in fielding reflect the rapid equipping, Capabilities Development for Rapid Transition, and other initiatives noted earlier.

The Army has concluded that shared situational awareness and unity of action are critical in current operations and that they can be improved by modernizing and expanding the Army's tactical network of radios, computers, digital devices, and sensors. The Army has, accordingly, renewed its efforts to enhance its network.[12] The Army responded to the growing toll from improvised explosive devices and mines by rapidly fielding mine-resistant, ambush-protected vehicles.

It is also regularly injecting enhanced capabilities into its units by incrementally fielding "capability packages." These packages are typically spin-offs from technologies that evolved within the Future Combat Systems program and are now suitable for integration with today's forces.

Task Organization

Task organization is one of a commander's principal tools for tailoring forces for the tasks or missions they are about to undertake. Task organization is usually a response to the commander's METT-TC analysis. That is, when given a new operations order, a commander considers what he has been directed to accomplish (the mission or "M" in METT-TC), the enemy's strength and disposition (the "E"), terrain and weather and the commander's own troops avail-

---

[10] Headquarters, U.S. Department of the Army, *Army Transformation: Report to the Congress of the United States, February 2007*, Washington, D.C., February 7, 2007.

[11] See U.S. Army, *2010 Army Posture Statement*, February 2010, p. 14.

[12] U.S. Army, 2010, p. 14.

able for the job (the "TT"), plus time and civil considerations (the "TC" in the acronym).[13] The commander then allocates forces on the basis of this analysis.[14] In the era of division-centric forces, division commanders typically task-organized their subordinate brigades by assigning them assets from the division base (e.g., engineers, aviation support) that were necessary for the brigades to accomplish their specific tasks in the operation. For example, if a brigade were to take the lead in breeching the enemy's defenses, it might be given additional engineering support to assist it. Task organization helped the commander align resources with tasks.

In the current force structure, many of the capabilities that once resided in the division base (military intelligence, engineers, signal, artillery, aviation, and medical, in particular) lie elsewhere in the force structure. BCTs have their own share, and some of these capabilities, like general support artillery, now reside in fires brigades. Similarly, aviation resides in one of three different models of combat aviation brigades. As a result, the frequency and echelon of task organization has changed slightly. In the current force structure, division and corps headquarters do not have organic resources to allocate to their subordinate units. Sustainment and functional support brigades must be assigned to them before they can task-organize any assets. BCTs typically require less augmentation from outside because their organic capabilities prepare them for most of the tasks they face. Former BCT commanders interviewed for this study indicated that their units were still task-organized, losing subordinate elements to other BCTs or gaining attachments from other BCTs, in addition to attachments from functional brigades. After-action reports note unmanned aerial vehicle sections, tactical human intelligence teams, explosive ordnance disposal teams, military working dogs, and signals intelligence detachments as typical attachments.

BCTs also tend to task-organize internally. Heavy BCTs may task-organize to create company teams that associate tanks with infantry. Another common practice is to task-organize to create more (sometimes up to seven) maneuver battalions from a BCT's organic units. This practice sometimes involves detaching a company from each maneuver battalion, dismounting all or part of the organic artillery battalion and recasting it as infantry, and retasking the RSTA squadron as infantry. Leadership for the new "battalions" is often drawn from among field-grade officers in the BCT headquarters.

It should be noted that the frequency, the echelon at which it takes place, and the extent of task organization should not be viewed as a symptom of faulty organizational design. Task organization is a response to METT-TC and therefore reflects the vast number of combinations and permutations of missions, enemy, weather, terrain, and civil considerations that can confront a military commander. Viewed in this light, task organization can be understood as a commander's attempt at creative problem-solving.

---

[13] U.S. Joint Chiefs of Staff, *Department of Defense Dictionary of Military and Associated Terms*, Joint Publication 1-02, Washington, D.C., 2001, as amended through September 2010.

[14] See HQDA, 2008.

Improvements and Pending Improvements in Brigade and Headquarters Designs

Most of the improvements undertaken thus far have added capabilities to the basic designs of current Army units.[15] These additions include truck platoons to enhance the tactical mobility of infantry BCTs and snipers to sharpen the lethality of heavy BCTs. The Army has created TDA augmentation packages that supplement MTO&E units by providing advisory assistance capabilities. The Army has also delivered engineer bridging to support units engaged in stability operations. It has also provided company-level intelligence teams by taking intelligence assets from echelons above brigade and reassigning them at the company level. Finally, the Army has also employed "in lieu of" deployment policies that substitute one type of unit for another. For example, an appropriately trained field artillery unit might be deployed in lieu of an MP formation if additional MP personnel were unavailable.

The Army has always sought to respond to feedback it receives from theater about the forces it provides: their organization, training, equipping, leadership, and so on. The fact that the Army makes these adjustments is probably just as indicative of the seriousness with which it treats its responsibilities under Title 10 of the U.S. Code to raise, train, equip, and maintain forces as it is of anything in particular about the current force structure. We would expect the Army to be just as diligent in responding to feedback from the field if it had deployed the previous, division-centric force structure. We would also expect similar amounts of feedback from the deployed forces and the regional COCOMs.

Deployability, Employability, and Sustainability of Modular Formations

For our purposes here, *deployability* refers to the amount of time needed to get a formation from its home station to the theater of operations. This definition is generally consistent with the definition in Joint Publication 1-02 although that definition emphasizes a unit's posture along a spectrum of readiness for deployment.[16] *Employability* here means the ease with which a unit can be brought into action upon its arrival in a theater of operations and its relative need for assistance in the process of reception, staging, and onward movement. *Sustainability* is "the ability to maintain the necessary level and duration of military activity to achieve military objectives."[17]

Deployability
The following judgments about deployability rest on earlier RAND analysis.[18] From that body of work, we conclude that the current force structure's deployability is roughly equivalent

---

[15] This discussion draws on interviews with ARCIC officials in September 2010.

[16] U.S. Joint Chiefs of Staff, 2010.

[17] U.S. Joint Chiefs of Staff, 2010.

[18] Alan J. Vick, David T. Orletsky, Bruce R. Pirnie, and Seth G. Jones, *The Stryker Brigade Combat Team: Rethinking Strategic Responsiveness and Assessing Deployment Options*, Santa Monica, Calif.: RAND Corporation, MR-1606-AF, 2002; Eric V. Larson, Derek Eaton, Paul Elrick, Theodore W. Karasik, Robert Klein, Sherrill Lingel, Brian Nichiporuk, Robert Uy, and John Zavadil, *Assuring Access in Key Strategic Regions: Toward a Long-Term Strategy*, Santa Monica, Calif.: RAND Corporation, MG-112-A, 2004; and Robert W. Button, John Gordon IV, Jessie Riposo, Irv Blickstein, and Peter A.

to that of its predecessor. There are marginal differences in loading and unloading times, although differences in the number and types of vehicles in the BCT types make somewhat different demands on airlift sorties and sealift. Deployability is also sensitive to the details in the scenario under consideration. For those that involve a road march to reach the initial area of operations, the modular BCTs' performance is superior because they have more vehicles and, thus, the road march requires less time in the overall calculation of deployability.[19]

Employability

A key phase in the employment of forces occurs when they arrive in the theater of operations. This phase includes the unit's reception, staging, and onward movement. A key consideration in assessing the comparative employability of units is the amount of assistance they need in disembarking, getting organized, locating their equipment, coordinating their movements out of the airfield or port, and using the road or rail infrastructure to reach their initial areas of operations.[20] During the Cold War, NATO earmarked thousands of "host-nation support troops" to facilitate the arrival of U.S. reinforcements in the event of a crisis during the Cold War.

We conclude that the current modular units are superior to their predecessors in terms of employability. This judgment is based on the fact that the modular BCTs and functional support brigades have greater capability in their headquarters to handle their own arrival, reception, staging, and onward movement. Their headquarters have better command and control than their predecessor units and are staffed with more senior NCOs and officers capable of handling the details of arrival and onward movement. This conclusion is also supported by the fact that the modular units have more organic transportation, CS, and CSS than their predecessors, which reduces dependencies on local support troops to receive the units in theater.

Sustainability

We estimate that the sustainability of the modular force is superior to that of its predecessor. This estimate is based, in part, on interviews conducted with members of the 311th Expeditionary Sustainment Command and 377th Transportation Battalion for another project. It also considers that today's units have enhanced command-and-control and RSTA systems, many of which are dependent on their host vehicles for power, that operate for extended periods ("long duty cycles," in Army parlance) and nevertheless remain functional. Most of the modular BCT designs are more equipment-intensive than their predecessors, yet their organic support assets manage to keep them operational.

Wilson, *Warfighting and Logistic Support of Joint Forces from the Joint Sea Base*, Santa Monica, Calif.: RAND Corporation, MG-649-NAVY, 2007.

[19] See Vick et al., 2002, Chapter Two.

[20] For a more detailed discussion, see Button et al., 2007, Appendixes A–C.

Bibliography

ARCIC—*see* U.S. Army Training and Doctrine Command, Army Capabilities Integration Center.

Edward L. Andrews, *The Army of Excellence and the Division Support Command*, Carlisle, Pa.: U.S. Army War College, May 21, 1986. As of November 17, 2010:
http://handle.dtic.mil/100.2/ADA168150

Belasco, Amy, *The Cost of Iraq, Afghanistan, and Other Global War on Terror Operations Since 9/11*, Washington, D.C.: Congressional Research Service, RL33110, September 28, 2009.

Bonds, Timothy M., Dave Baiocchi, and Lauri L. McDonald, *Army Deployments to OIF and OEF*, Santa Monica, Calif.: RAND Corporation, DB-587-A, 2010. As of November 17, 2010:
http://www.rand.org/pubs/documented_briefings/DB587/

Button, Robert W., John Gordon IV, Jessie Riposo, Irv Blickstein, and Peter A. Wilson, *Warfighting and Logistic Support of Joint Forces from the Joint Sea Base*, Santa Monica, Calif.: RAND Corporation, MG-649-NAVY, 2007. As of November 17, 2010:
http://www.rand.org/pubs/monographs/MG649/

Cianciolo, Mark G., *U.S. Army Strike Force—A Relevant Concept?* Ft. Leavenworth, Kan.: School of Advanced Military Studies, April 1999.

Condon, R. L., "TF Modularity: Answers to CSA Questions, Pre-Brief to MG Mixon 20 August 2004," briefing, U.S. Army TRADOC Analysis Center, 2004.

Congressional Budget Office, *An Analysis of the Army's Transformation Programs and Possible Alternatives*, Washington, D.C., No. 3193, June 2009.

Congressional Research Service, *U.S. Army's Modular Redesign: Issues for Congress*, Washington, D.C., RL32476, updated May 5, 2006.

Davis, Lynn E., J. Michael Polich, William H. Hix, Michael D. Greenberg, Stephen Brady, and Ronald E. Sortor, *Stretched Thin: Army Forces for Sustained Operations*, Santa Monica, Calif.: RAND Corporation, MG-362-A, 2005. As of November 17, 2010:
http://www.rand.org/pubs/monographs/MG362/

DoD—*see* U.S. Department of Defense.

Donnelly, William M., *Transforming an Army at War: Designing the Modular Force, 1991–2005*, Washington, D.C.: U.S. Army Center of Military History, 2007.

Green, Wayne A., *Interim Strike Force Headquarters Digital LNO Nodes: Force Tailoring Enablers*, Ft. Leavenworth, Kan.: Army Command and General Staff College, May 1999.

Headquarters, U.S. Department of the Army, *Operations*, Field Manual 100-5, Washington, D.C., June 14, 1993.

———, "Answers to CSA Questions," task force modularity briefing, August 20, 2004.

———, *Army Transformation: Report to the Congress of the United States, February 2007*, Washington, D.C., February 7, 2007.

———, *Operations*, Field Manual 3-0, Washington, D.C., February 27, 2008.

Headquarters, U.S. Department of the Army, and U.S. Marine Corps, *Operations Terms and Symbols*, Field Manual 101-5-1 and Marine Corps Reference Publication 5-2A, Washington D.C., September 30, 1997.

HQDA—*see* Headquarters, U.S. Department of the Army.

Joint Forces Staff College, *The Joint Staff Officer's Guide 2000*, Publication 1, Norfolk, Va., 2000. As of November 17, 2010:
http://handle.dtic.mil/100.2/ADA403118

Larson, Eric V., Derek Eaton, Paul Elrick, Theodore W. Karasik, Robert Klein, Sherrill Lingel, Brian Nichiporuk, Robert Uy, and John Zavadil, *Assuring Access in Key Strategic Regions: Toward a Long-Term Strategy*, Santa Monica, Calif.: RAND Corporation, MG-112-A, 2004. As of November 17, 2010:
http://www.rand.org/pubs/monographs/MG112/

Layton, Brian, Office of the Deputy Chief of Staff G-8, U.S. Army, *Former Brigade Commander Seminar #2*, May 13, 2009.

Linick, Michael E., *A Critical Evaluation of Modularity*, Carlisle, Pa.: U.S. Army War College, March 8, 2006.

Nardulli, Bruce, Walter L. Perry, Bruce R. Pirnie, John Gordon IV, and John G. McGinn, *Disjointed War: Military Operations in Kosovo, 1999*, Santa Monica, Calif.: RAND Corporation, MR-1406-A, 2002. As of November 17, 2010:
http://www.rand.org/pubs/monograph_reports/MR1406/

Office of the Secretary of Defense, *Conduct of the Persian Gulf War, Final Report to Congress Pursuant to Title V of the Persian Gulf Conflict Supplemental Authorization and Personnel Benefits Act of 1991 (Public Law 102-25)*, Washington, D.C., April 1992.

Perry, Walter L., and Marc Dean Millot, *Issues from the 1997 Army After Next Winter Wargame*, Santa Monica, Calif.: RAND Corporation, MR-988-A, 1998. As of November 17, 2010:
http://www.rand.org/pubs/monograph_reports/MR988/

Pickup, Sharon, and Janet St. Laurent, Defense Capabilities and Management, U.S. Government Accountability Office, *Force Structure: Preliminary Observations on Army Plans to Implement and Fund Modular Forces*, testimony before the Subcommittee on Tactical Air and Land Forces, House Armed Services Committee, Washington, D.C., GAO-05-443T, March 16, 2005.

Public Law 111-84, National Defense Authorization Act for Fiscal Year 2010, October 28, 2009.

Quinlivan, James T., "Force Requirements in Stability Operations," *Parameters*, Vol. 25, No. 4, Winter 1995, pp. 59–69.

Reardon, Mark J., and Jeffery A. Charlston, *From Transformation to Combat: The First Stryker Brigade at War*, Washington D.C.: U.S. Army Center of Military History, 2007.

Rostker, Bernard D., *I Want You! The Evolution of the All-Volunteer Force*, Santa Monica, Calif.: RAND Corporation, MG-265-RC, 2006. As of November 17, 2010:
http://www.rand.org/pubs/monographs/MG265/

Schoomaker, General Peter J., U.S. Army Chief of Staff, testimony before the Senate Armed Services Committee, February 10, 2004a.

———, U.S. Army Chief of Staff, prepared statement before the House Armed Service Committee, July 21, 2004b.

Shinseki, General Eric K., U.S. Army Chief of Staff, prepared statement before the Senate Armed Services Committee, October 26, 1999.

———, U.S. Army Chief of Staff, statement before the Senate Armed Services Committee, March 1, 2000.

TRADOC—*see* U.S. Army Training and Doctrine Command.

U.S. Army, *2004 Posture Statement*, February 5, 2004. As of November 17, 2010:
http://www.army.mil/aps/04/

———, *2010 Army Posture Statement*, February 2010. As of November 17, 2010:
https://secureweb2.hqda.pentagon.mil/vdas%5Farmyposturestatement/2010

U.S. Army Training and Doctrine Command, *Force XXI Operations*, Pamphlet 525-5, Fort Monroe, Va., August 1, 1994.

———, *Concepts of Modularity*, Pamphlet 525-68, Fort Monroe, Va., January 10, 1995.

———, *ARCIC Comprehensive Lessons Learned*, August 2009.

U.S. Army Training and Doctrine Command, Army Capabilities Integration Center, "Overarching Modular Briefing to CSA Backups v 1.2 011500," February 2004.

U.S. Department of Defense, *Joint Strategic Capabilities Plan 2010*, 2010a.

———, *Quadrennial Defense Review Report*, Washington, D.C., February 2010b.

U.S. Government Accountability Office, *Army Needs to Provide DoD and Congress More Visibility Regarding Modular Force Capabilities and Implementation Plans*, Washington, D.C., GAO-06-745, September 2006.

U.S. Joint Chiefs of Staff, *Department of Defense Dictionary of Military and Associated Terms*, Joint Publication 1-02, Washington, D.C., 2001, as amended through September 2010.

Vick, Alan J., David T. Orletsky, Bruce R. Pirnie, and Seth G. Jones, *The Stryker Brigade Combat Team: Rethinking Strategic Responsiveness and Assessing Deployment Options*, Santa Monica, Calif.: RAND Corporation, MR-1606-AF, 2002. As of November 17, 2010: http://www.rand.org/pubs/monograph_reports/MR1606/

Watson, Brian G., *Reshaping the Expeditionary Army to Win Decisively: The Case for Greater Stabilization Capacity in the Modular Force*, Carlisle, Pa.: U.S. Army War College, August 2005.

Wilson, John B., *Maneuver and Firepower: The Evolution of Divisions and Separate Brigades*, Washington, D.C.: U.S. Army Center of Military History, 1998.